Bibliographic information published by the Deutsche Nationalbibliothek

The Deutsche Nationalbibliothek lists this publication
in the Deutsche Nationalbibliografie; detailed bibliographic data
are available in the Internet at http://dnb.d-nb.de .

ISBN 978-3-8325-2100-4

Logos Verlag Berlin GmbH
Comeniushof, Gubener Str. 47,
D-10243 Berlin
Germany
Tel.: +49 030 42 85 10 90
Fax: +49 030 42 85 10 92
INTERNET: http://www.logos-verlag.de

Nelson Baloian, Wolfram Luther, Dirk Söffker, Yoshiyori Urano (Eds.)

Multimodal Human-Machine Interaction in Different Application Scenarios

International DAAD-PhD Summer Academy at the University of Chile, Santiago de Chile, August 27 to September 10, 2008

Revised Contributions

Volume Editors

Nelson Baloian
Department of Computer Science – Universidad de Chile
Blanco Encalada 2120, Santiago 6511224, Chile
E-mail: nbaloian@dcc.uchile.cl

Wolfram Luther
University of Duisburg-Essen,
Chair of Scientific Computing, Computer Graphics, and Image Processing
Lotharstraße 65, 47048 Duisburg, Germany
E-mail: luther@inf.uni-due.de

Dirk Söffker
University of Duisburg-Essen,
Chair of Dynamics and Control
Lotharstraße 1, 47048 Duisburg, Germany
E-mail: dirk.soeffker@uni-due.de

Yoshiyori Urano
Graduate School of Global Information and Telecommunication Studies
Waseda University
Honjo-Campus, Japan
E-mail: urano@waseda.jp

ACM Subject Classification (1998): H.1.2, H.5.2

Preface

The goal of this first of three DAAD Summer Academies located in Santiago de Chile was to bring together highly motivated young researchers at the PhD level with their professors and internationally renown experts in the emerging interdisciplinary fields of human-machine interaction and computer supported cooperative work. Additionally, the organizers hoped to extend the bi-national PhD-net programme recently launched by the German Academic Exchange Service (DAAD) to the three Universities of Chile, Duisburg-Essen, Germany, and Waseda in Tokyo, Japan.

During a two-week period at the end of August 2008, twenty-six participants took part in the DAAD Summer Academy organized by the DCC (Departamento de Ciencias de la Computación) at Santiago de Chile. This publication gathers the original material presented by the lecturers along with brief introductions and explanations. This collection shows the state of the art in the fields of human-machine interactions in different application scenarios, modeling and evaluating groupware systems and modern optical communication systems, including reviews of the development of the scientific fields involved, and new ideas and research directions. The Human-Machine Interaction in Networked Systems series will be continued at the University of Duisburg-Essen in summer 2009 focusing on interface and interaction design for learning and simulation environments and at the University of Waseda working on collaborative learning interactions in communities of practice and learning in springtime 2010.

CONTENTS

COMMUNICATION SYSTEMS

Acknowledgments

The organizers of the international PhD Summer Academy 2008 in
Santiago de Chile would like to express their deep gratitude to all
participating lecturers. Without their contributions neither the Summer Academy nor this book would have been possible. Special
thanks is due to the German Academic Exchange Service (DAAD)
for funding three consecutive Summer Academies in Santiago, Duisburg, and Tokyo.

Evaluating Collaborative Work

José A. Pino, Valeria Herskovic
Department of Computer Science, University of Chile

Abstract

The evaluation of collaborative systems is an important yet not fully solved problem in the field of Computer-Supported Cooperative Work (CSCW). There are many proposed methods, but they have not been widely used: many groupware systems are deficiently evaluated.

The first part of this work will summarize a selection of twelve main groupware evaluation methods. They will be classified according to various criteria. Furthermore, a strategy to find the most appropriate method will be suggested.

The second part of the work will concentrate on formal evaluation methods. In particular, two of the above methods will be described in detail. The first one, called Performance Analysis (PAN) models collaborative work as a task to be performed by a number of people in a number of stages, and the concepts of quality, time and total amount of work done are defined; the evaluators must define a way to compute the quality and maximize the quality vs. work done. The second formal method to be studied is the Human Performance Method (HPM). This method adapts the keystroke level method (KLM) to a group of users communicating through a shared workspace. In this method, evaluators first decompose the physical interface into several shared workspaces. Then, they define critical scenarios focused on the collaborative actions for the shared workspaces. Finally, evaluators compare group performance in the critical scenarios to predict execution times.

The third part of the work will focus on a proposal to evaluate mobile groupware. First, a characterization of loosely coupled mobile collaboration (LCMC) will be presented, along with three case studies: firefighters dealing with search and rescue scenarios home care health professionals, and building construction inspections. An analysis technique for LCMC will be shown, and in combination with a list of common requirements for mobile shared workspaces, will permit the definition of a new evaluation method for this type of software.

Part I : Groupware Evaluation Methods
Contents
- Groupware
- Why do we need to evaluate Groupware?
- Representative evaluation methods
- Selecting an evaluation method
- Organizing the evaluation process
- Conclusions

1. Groupware
- Software designed to help people involved in a common task achieve their goals
- Basis for Computer Supported Collaborative Work (CSCW)
- Small groups – large groups

- Human factors are important
- People and MIS: "People must help the system perform well" (entering data…)
- People and CSCW: "The system helps people achieve their goals" (supporting needs…)
- Much software oriented to support individual tasks
- However, little support for group work or team work
- E.g., up to 70% of people's time working in offices is spent in meetings
- Office automation? Not oriented to clerical personnel! Rather: professionals, executives…

2. The 3-C model
- Communication
- Collaboration
- Coordination
- Varying degrees of each ingredient

3. Communication tools
Messages, files, data, documents
- Synchronous conferencing
- Wikis
- E-mail
- Web publishing

4. Conferencing tools
Sharing information in interactive way
- Internet forums
- Data conferencing (whiteboards)
- Videoconferencing
- Application sharing (e.g., a workspace)
- Electronic meeting systems

5. Collaborative management tools
Supporting group activities
- Electronic calendars
- Workflow systems
- Knowledge management systems
- Online spreadsheets
- Social software systems

6. Groupware systems classification
[Ellis et al., 1991]

Time Space	Synchronous	Asynchronous
Co-located		
Distributed		

7. Why do we need to evaluate Groupware?

- Few groupware systems are evaluated
- Causes
- Single user evaluation is not applicable
- Expensive
- Unavailable resources
- Long-term
- Which method to use?

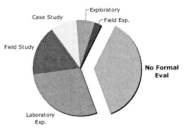

[Pinelle and Gutwin, 2002]

8. Motivation: How to evaluate?

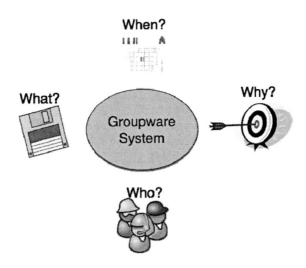

9. Evaluation Methods
- Summative Evaluation
- Formative Evaluation

Some methods are adaptations from single user methods…
Some methods are applicable to certains kinds of groupware only

10. Groupware Heuristic Evaluation (GHE)
- Adaptation of Heuristic Evaluation (HE)
- Usability experts check UI complies with usability principles (heuristics)
- Eight heuristics inspired by Mechanics of Collab.
- Evaluators individually study the UI
- Reported problems are consolidated
- The list serves to improve the system and correct errors during or after development

11. Groupware Walkthrough (GWA)
- Based on Cognitive Walkthrough (CW)
- CW: task sequences to explore users' actions
- Group tasks, multiple users, concurrency? GWA!
- Scenario: activity, users, knowledge, outcome
- Scenario building by users' observation at work
- Evaluators walk through tasks in laboratory
- Meeting to analyze results
- Formative. Not a replacement of field methods!

12. Collaboration Usability Analysis (CUA)
- Task analysis technique with focus on teamwork aspects of collaboration in shared tasks
- Collaborative action $\Rightarrow$ Collab. Mechanisms $\Rightarrow$ UI elements
- Diagrams: task components, flow through them, distribution of tasks to people
- Variable paths on the execution of a task

13. Groupware Observational User Testing (GOT)
- Based on Observational User Testing (OUT)
- Task performance observation in lab
- Monitoring users having problems with a task, or ask users to think aloud about their work
- Focus on collaboration aspects
- Predefined criteria (mechanics of collaboration)
- System prototype or finished system

14. Human-Performance Models (HPM)
- Low-level model of interaction with UI
- Laboratory setting

- No users, no prototypes
- What if? analysis
- Formative or summative

15. "Quick-and-dirty" Ethnography (QDE)
- Ethnography: Qualitative observation of human social phenomena $\Rightarrow$ detailed descriptions
- Prolonged activity
- Access to relevant sites and people
- QDE: brief workplace studies
- Broad understanding
- Deficiencies are captured $\Rightarrow$ acceptability, usability
- Improvement current systems, inform design

16. Performance Analysis (PAN)
- Quantitative method
- Based on relationships between output quality, time spent, total work done
- No access to users or prototypes

17. Perceived Value (PVA)
- Measuring the perceived value of meetingware by its users
- System contributions to indiv., group & organiz.
- developers identify relevant components
- external attributes to be evaluated
- after use, fill out evaluation map

18. Scenario-Based Evaluation (SBE)
- Provide realistic setting to perform eval.
- Scenario: concrete detailed activity description (actor, context, task goals, claims)
- Evaluation: eliciting claims from participants
- Case: semi-structured interviews: background
- Focus groups to validate, clarify results
- Frequency analysis: identify key scenarios
- Redesign from positive and negative claims

19. Cooperation Scenarios (COS)
- Textual descriptions of work practices, including motivation and goals
- Field studies, semi-structured interviews, visits
- First part is without users: new design is analyzed to check changes and who benefits
- Second part is with actual users, with a prototype

20. E-MAGINE (EMA)
- First: profile of group and groupware application
- Second: evaluation and feedback

- Meeting with client: goals
- Semi-structured interview with a group member
- CSF, e.g., application robustness, motivation
- Issues, e.g., social cohesion, usability
- Instruments, e.g., Know. Sharing Sup. Inventory
- Results communicated to the group

21. Knowledge Management Approach (KMA)
- It measures whether the system helps users detect knowledge flows and disseminate, store and reuse knowledge
- Areas: creation, accumulation, sharing, utilization, internalization
- Questionnaires

22. Other techniques
- Logging
- Questionnaires
- Interviews
- Focus groups
- Video recordings

23. Selecting an Evaluation Method
- Identification of evaluation scenario
- Stakeholder chooses method by key features
- After first filtering
  - If no suitable methods exist, design ad hoc method
  - If several are found, choose according to prioritization of characteristics

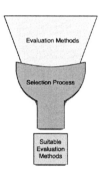

24. Selecting an Evaluation Method

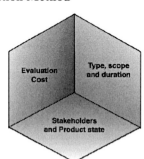

25. Stakeholders / Product state

	Developers	Users	Manager
Products Under Development	GHE, GWA CUA, HPM, PAN	SBE, COS EMA, KMA	PVA
Finished Products		GOT, QDE	GOT, QDE PAN, PVA

- Fast to instantiate
- Highly relevant

26. Evaluation Method Features (1)

Evaluation Method	Who			When		
	Users	Develop.	Experts	Before	Summ.	Form.
GHE			●		●	●
GWA					●	●
CUA					●	●
GOT	●				●	●
HPM				●	●	●
QDE	●				●	
PAN					●	●
PVA	●	●		●	●	●
SBE	●				●	●
COS	●					●
EMA	●				●	●
KMA					●	●

27. Evaluation Method Features (2)

Evaluation Method	Type		Location	
	Quantitative	Qualitative	Workplace	Laboratory
GHE		●		●
GWA		●		●
CUA		●		●
GOT		●		●
HPM	●			●
QDE		●	●	
PAN	●			●
PVA		●	●	
SBE		●	●	
COS		●	●	
EMA		●	●	
KMA		●	●	

28. Evaluation Method Features (3)

Evaluation Method	Time Span			Goal		
	Hours	Days	Weeks	Product	Process	Context
GHE	●			●		
GWA		●			●	
CUA		●			●	
GOT			●		●	
HPM		●			●	
QDE			●			●
PAN		●		●		
PVA		●		●		
SBE			●		●	●
COS			●		●	●
EMA		●			●	●
KMA		●		●		●

29. Evaluation Method Features

Evaluation Method	Who			When			Type		Location		TimeSpan			Goal		
	Users	Develop.	Experts	Before	Summ.	Form.	Quantitative	Qualitative	Workplace	Laboratory	Hours	Days	Weeks	Product	Process	Context
GHE			●		●	●		●		●	●			●		
GWA					●	●		●		●		●			●	
CUA					●	●		●		●		●			●	
GOT	●				●	●		●		●			●		●	
HPM				●	●	●	●			●		●			●	
QDE	●				●	●		●	●				●			●
PAN					●	●	●	●		●		●		●		
PVA	●	●		●	●	●		●		●		●		●		
SBE	●				●	●		●		●			●		●	●
COS	●				●	●		●		●			●		●	●
EMA	●				●	●		●	●	●		●			●	●
KMA					●	●		●		●		●		●		●

30. Evaluation Cost Classification

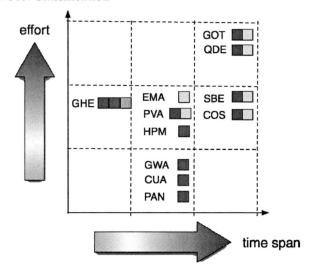

31. Application Example (1)

- **Developers** need to evaluate a brainstorming tool to **find problems in collaboration process.**

	Developers
Products Under Development	GHE, GWA CUA, HPM, PAN

- Classification criteria: developers, process

	Users	Develop.	Experts		Product	Process	Context
GHE			●	PAN	◯		

32. Application Example (2)

- Suitable methods: GWA, CUA, HPM

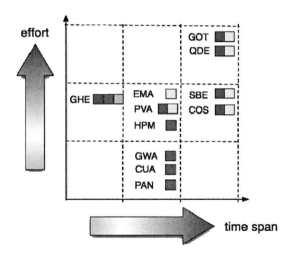

- Minimize effort: choose GWA or CUA

33. Organizing the Evaluation Process

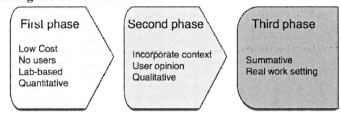

First phase

Low Cost
No users
Lab-based
Quantitative

Second phase

Incorporate context
User opinion
Qualitative

Third phase

Summative
Real work setting

34. Conclusions
- Unevaluated groupware may be unsuccessful
- Classifications provide fast comparison
- Help choose evaluation method
- Some areas lack evaluation methods
- Apply several evaluation methods --> comprehensive understanding

35. Future Work
- Test usability of classifications
- Improve classification criteria
- Provide application to aid in selection of evaluation method

36. Bibliography of Part I
- C.A. Ellis, S.J. Gibbs, G.L. Rein: Groupware – some issues and experiences. *Communications of the ACM* 34(1), 1991, 39-58.
- V. Herskovic, J.A. Pino, S.F. Ochoa, P. Antunes: Evaluation methods for Groupware. *Lecture Notes in Computer Science* 4715, 2007, 328-336.

Part II : Formal Evaluation Methods
Contents:
- Towards Formal Evaluation
- Collaborative Retrieval
- Performance Analysis
- Collaborative Workspace Design
- Human Performance Analysis

1. Collaborative Retrieval
- Information Retrieval
- "Get all the stored relevant information concerning X, and nothing else"
- X: concept, not words
- Relevant? $\Rightarrow$ Precision

2. Information Retrieval
- Precision: proportion of relevant retrieved material over all retrieved material
- Recall: proportion of relevant retrieved material over all relevant material

3. Information Retrieval

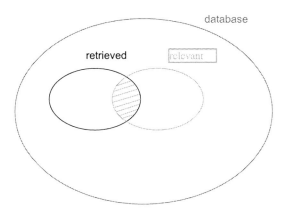

4. Information Retrieval

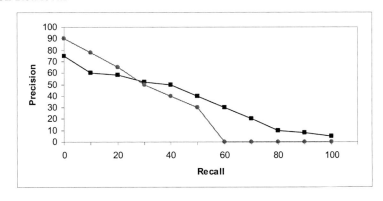

5. Information Retrieval
- How to increase Recall?
- Example: legal information retrieval: car accidents [Blair and Maron, 1985]
- Several ways: e.g., thesaurus
- We propose collaborative retrieval

6. Collaborative Retrieval

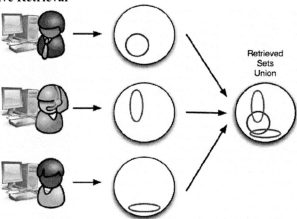

Retrieved
Sets
Union

7. Bates' experiment (1998)
- Freedom of speech + Internet
- First ammendment + Web
- Free speech + Cyberspace
- Intellectual freedom + Net
- AltaVista search engine

8. Bates' experiment (cont'd)
- Ten first retrievals of each query have no intersection
- Combining terms: 16 queries, 160 retrievals
- 138 unique retrievals

9. PAN Evaluation method
- Assumptions
- Quality of a project can be assessed
- Group of m persons
- Same abilities and task performance
- Task divided into n stages

10. Stages
- Time-based (e.g., days)
- Location-based (e.g., meetings)
- Task-based (e.g., sub-projects)

11. Efficiency measures
- Quality: how good is the result of the collaborative work?
- Time: Total elapsed time (parallelism)
- Work: Total amount of work done (with parallel work: (time spent) * m)

12. Simple analysis
- In most cases, Quality improvement decreases with every stage

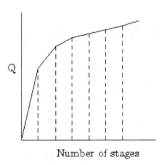

13. Simple analysis (cont'd)
- Quality vs. number of people

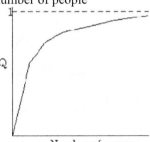

14. Simple analysis (cont'd)
- Typically, work increases with the number of people at a rate higher than linear

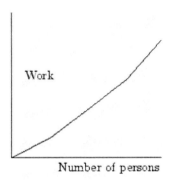

15. Maximizing quality per work done
- Combining previous functions

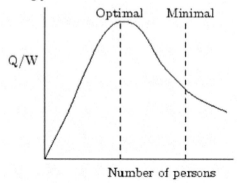

16. Example
- $Q = 1 - e^{-m/\alpha}$
- (α measures how fast the collaborative task is saturated by m people)
- Maximizing Q/m:
- Approximate solution: $m \approx 3\alpha/2$

17. Time optimality
- Parallelism saturates due to superlinear interaction time

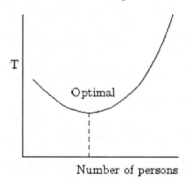

18. Collaborative Retrieval analyzed
- The simple method described before may generate high recall, but low precision
- Precision and recall may be difficult to estimate
- Another strategy: Union-Refine (U-R)

19. Union-Refine

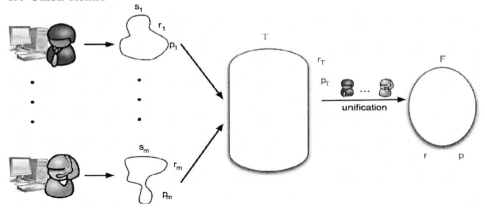

20. U-R analyzed

- P1. $p_T \leq max_i \{p_i\}$
- P2. $p \geq p_T$
- P3. For all i, $r_T \geq r_i$
- P4. $r \leq r_T$ (almost the same)

21. U-R analyzed (cont'd)

- Measures of quality:
- Group precision: $GP = p/max_i \{p_i\}$
- Group recall: $GR = r/max_i \{r_i\}$
- $GP \leq 1$ and $GR \geq 1$
- The bigger, the better collaboration performance over individual contributions
- $Q = GP$ or $Q = GR$

22. U-R analyzed (cont'd)

- Example: quality per work done defined as
- $Q / f(m)$
- $f(m)$: financial cost of having *m* people searching
- Optimal number of people: maximize $Q / f(m)$
- Analytical or experimental analysis

23. Collaborative Workspace Design

24. Collaborative Workspace Design (cont'd)
- Optimize shared workspace performance
- Measuring performance is typically expensive
- Discount methods (qualitative)
- Our approach: quantitative predict and compare
- GOMS [Card et al., 1983]: Goals, Operators, Methods and Selection Rules
- Analysis and modeling techniques, many studies on diverse applications

25. Model Human Processor (MHP)

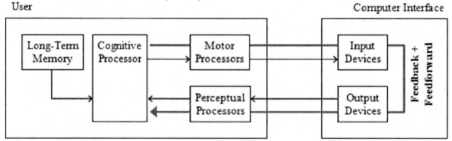

26. MHP (cont'd)
- Application to groupware [Kieras & Santoro, 2004]
- Interaction *between* users
- Information flows should consider collaboration, mutual awareness, interdependence.

27. MHP (cont'd)
- *Feedback*: info to make user aware of executed operations
- *Feedforward*: initiated by the UI to make user aware of available action possibilities
- For groupware, we need additional categories

28. MHP (cont'd)
- *Explicit communication*: one user to another one
- *Feedthrough*: implicit info delivered to several users reporting actions executed by one user (group awareness and meaningful contexts)

- *Back-channel feedback:* unintentional info initiated by one user directed to another one, indicating the listener is following the speaker [Rajan et al., 2001]. No meaningful content

29. Groupware information flows

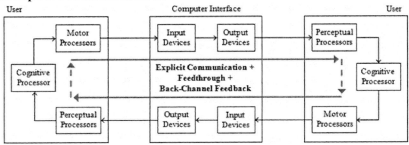

30. Groupware specializations

- Awareness input/output devices:
    - Who, what, when, how, where are the other users
    - Allow users to perceive role and limitations of the UI as mediator
    - Loose the link between executed operations and group awareness
    - Coupling at origin and at destination

31. Groupware interface with specialized devices

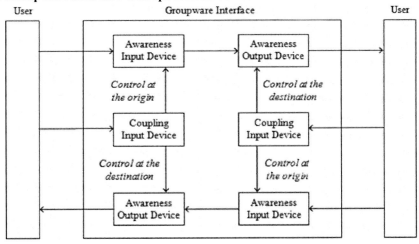

32. The HPM Method
- Step 1. *Groupware Interface* (awareness and coupling devices)
- Step 2. *Critical Scenarios* (collab. actions with potential important effects on performance)
- Step 3. *Boundary selection* (focus)
- Step 4. *SW performance* (predicted exceution time)

33. KLM (Keystroke Level Model)
- Each user action is converted to a sequence of mental and motor operators [Card et al., 1980]
- Each operator has been empirically established and psychologically validated [Kieras, 2003] [Olson & Olson, 1990]

34. KLM (cont'd)
H	0.4	Home hand on keyboard
K	0.1	Press or release mouse button
M	1.2	Mentally prepare
P	1.1	Point with mouse to target on display

35. Example of Use
- Reserving objects in a WS
- Object can be changed only by user who reserved it
- Users must be aware of reserved objects
- When reserving, one of two outcomes: success or failure
- Goal: minimize time wasted on simultaneous reservations for same object

36. Workspaces

Shared

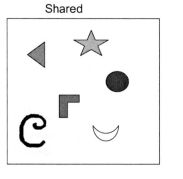

Private X

Private Y

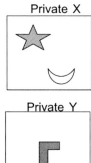

37. Step 1: Groupware Interface
- Scenario A: No awareness
- Scenario B: after successful reservation, letter identifying current owner
- Scenario C: while user is selecting objects, a rectangle comprising objects is shown
- (feedthrough, loose coupling)

38. Step 2: Critical scenarios
- It happens when users try to reserve same object in parallel
- A: users behave as if all objects are available
- B: letters are shown only *after* reservation
- C: besides letters, rectangles; users will likely choose other objects to work with

39. Step 3: Boundary selection
- All objects in the WS are visible
- Feedthrough is instantaneous
- It is unlikely that more than *two* users select the same object at the same time
- first user entering a competition for the same objects always succeeds in making the reservation

40. Step 4: Shared WS performance
- Sequence of KLM operators is the same
- Difference for three scenarios is caused by availability and timeliness of awareness information
- First search for objects: M
- Then, move mouse pointer: P
- Press mouse button: K
- Total: MPKPKPKPK: 6 seconds

41. Step 4: Shared WS performance

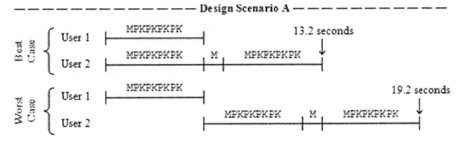

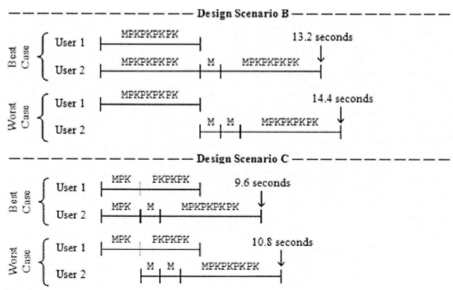

42. Other examples
- Locating updated objects
- Negotiating requirements (SQFD)
- Students' attention during face-to-face lecturing

43. Conclusions
- Both methods provide quantitative evaluations
- Low-cost
- Formative
- Attention to meaningful details
- They could be extended to other cases/applications

44. Bibliography of Part II
- A. Ferreira, P. Antunes, J. A. Pino : Evaluating Shared Workspace Performance using Human Information Processing Models. *Information Research* 2008, *to appear*.
- N. Baloian, J. A. Pino, H. U. Hoppe: Dealing with the students' attention problem in computer supported face-to-face lecturing. *Educational Technology and Society* 11(2), 2008, 192-205.
- R. A. Baeza-Yates, J. A. Pino: Towards Formal Evaluation of Collaborative Work and its application to Information Retrieval. *Information Research* 11(4), paper 271, 2006.

Part III: Evaluation of Mobile Shared Workspaces to Improve their Support for Collaboration

Contents:
- Mobile Groupware Evaluation
- Loosely coupled mobile collaboration
- Modeling mobile collaboration

1. Mobile Shared Workspaces (MSW)
- Persistent space, synchronous or asynchronous interaction [Dourish and Bellotti, 92]
- Mobility should not hinder cooperation [Divitini et al., 2004]
- Challenges from collaboration and mobility

2. Motivation
- Groupware Evaluation
  - Complexity, cost, required resources.
  - Single user evaluation is not directly applicable.
  - Lack of global perspective.

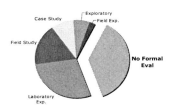

[Pinelle and Gutwin, 2002]

3. Motivation
- Mobile app evaluation
  - Problems with setting
- Mobile groupware evaluation [Kellar et al., 2003]
  - Finding widely deployed app
  - Context
  - Task-centric
  - Verbal data

4. Problem Description
- Lack of effective, low cost evaluation methods.
- Developers do not evaluate intermediate products.
- Final product lacks context and may be unsuccessful.
- Our goal: to define a formative evaluation method to help MSW developers improve support for **collaboration**.

5. Evaluation Process

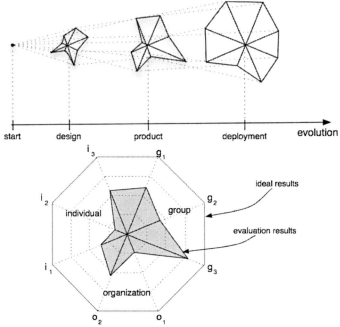

6. Mobile Collaboration Examples
- Search & Rescue
- Construction
- Home health care

7. Emergency Management Case Study

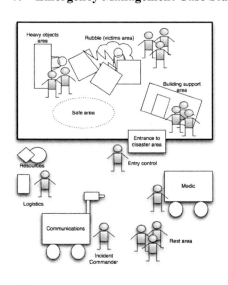

8. **Stages of Work**
    -   Size up (Reconnaisance and Organization)

702   704   708

706

Cuadra 700 Calle Alfa

701

703   705   707

9. **Stages of Work**
    -   Search
        -   Find highest number of victims as fast as possible
        -   Interviews and systematic search
        -   Marking system
    -   Rescue
        -   Resource allocation for successful rescue

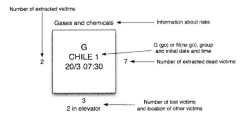

10. **Collaboration Support**
    -   Radio
        -   Resource and personnel allocation
        -   Notifications
        -   Reports
    -   Marking systems
        -   Communication of status
    -   Acoustic alerts
    -   Face to face

11. Construction Inspections
- Workspace with construction blueprint
- Users annotate with problems, comments, etc
- Data is shared and reviewed

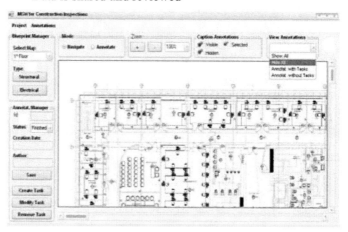

12. Home Care Services [Pinelle, 2004]
- Patient receives services at home, from up to 7 disciplines.

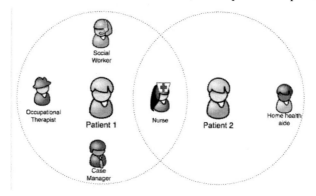

13. Loosely Coupled Mobile Collaboration (LCMC)
- **LCMC** refers to the work of a **group** of people working in **weakly interdependent tasks** in which the group is **on the move** and distributed in space, time, or both.
- To support this type of work technologically, **mobile** devices (such as cell phones, PDAs, tablet PCs, laptops, etc) should be used.

14. LCMC: Autonomous and collaborative work

- In LCMC, work is autonomous, with irregular and unpredictable periods of collaboration.

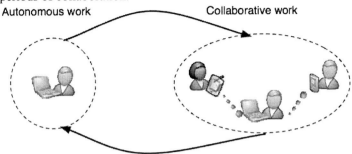

Autonomous work   Collaborative work

15. CSCW Time/Space Matrix

	Same place	Different places
Same time	**Face to face** •GDSS •Public displays	**Remote interaction** •Video conferencing •Media spaces
Different time	**Ongoing tasks** •Project management •Group displays	**Comm., coordination** •Discussion board •E-mail

[Johansen, 1988]

- In mobile collaboration: what does *space* mean?

16. LCMC: Simultaneity/Reachability

- **Simultaneous:** Available to work synchronously
- **Non simultaneous:** Asynchronous work
- **Reachable:** Exchange information in highly predictable way

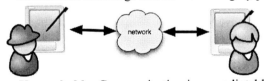

- **Unreachable:** Communication is unpredictable, or there is no communication

17. LCMC: Simultaneity/Reachability Matrix

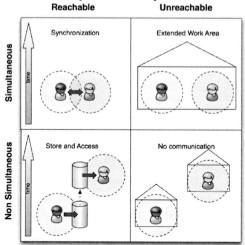

18. LCMC: Transitions in Matrix

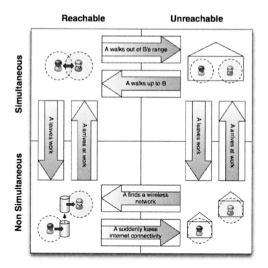

19. Modeling LCMC

- Graph G=(N,A)
- Nodes
  - o Role present in collaboration
  - o Intermediary, permits collaboration
- Arc: collaboration between two roles

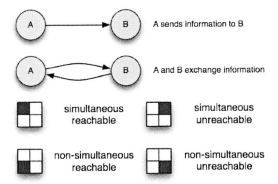

20. Example: MSW for construction inspections

- HQ where project leader and staff work
- Construction site where foreman, group leader, inspectors work.
- Data is sent to HQ or shared on site.

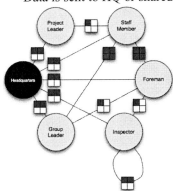

21. General Requirements for MSW

- Flexibility
  - User connection/disconnection
  - Automatic peer detection
- Consistency and Availability
  - Explicit data replication
  - Caching
  - Conflict resolution
- Connectivity
  - Automatic connection
  - Service and device discovery
  - Message routing
  - User gossip

22. General Requirements for MSW

- Heterogeneity and Interoperability
  - Heterogeneity

        o   Interoperability
- Communication
  - o   Synchronous messaging
  - o   Asynchronous messaging
  - o   File transfer
  - o   Pushing notifications

23. General Requirements for MSW

- Awareness
  - o   Online awareness
  - o   Offline awareness
  - o   Transition awareness
- Protection
  - o   Ad-hoc work sessions
  - o   Security
  - o   Privacy

24. Requirements

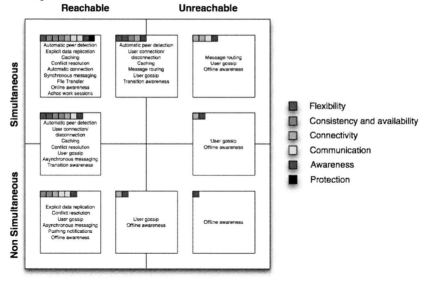

25. Evaluation method proposal

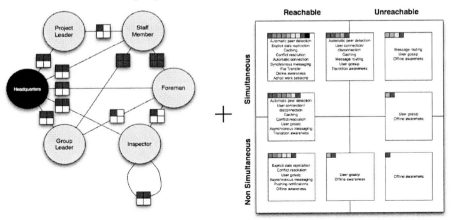

- Automatic test generation

26. Conclusions
- **Formative** evaluation method
- No **usability** evaluation, only collaboration support evaluation
- For **developers** (no users or experts)
- Future Work
  - Software tool for **graph creation** and **automatic analysis**
  - Evaluation method **assessment**

27. References of Part III

- M. Divitini, B. A. Farshchian, and H. Samset. Ubicollab: collaboration support for mobile users. In *SAC '04: Proceedings of the 2004 ACM symposium on Applied computing*, pages 1191–1195. ACM Press, 2004.
- P. Dourish and V. Bellotti. Awareness and coordination in shared workspaces. In *Proceedings of CSCW'02*, pp. 107–114. ACM Press, 1992.
- V. Herskovic, S. Ochoa, J. Pino, A. Neyem, General Requirements to Design Mobile Shared Workspaces, *Proceedings of CSCWD'08.*
- M. Kellar, K. Inkpen, D. Dearman, K. Hawkey, V. Ha, J. MacInnes, B. MacKay, M. Nunes, K. Parker, D. Reilly, M. Rodgers, T. Whalen, Evaluation of Mobile Collaboration: Learning from our Mistakes, Technical Report CS-2004-13, Dalhousie University, Canada, 2004.
- R. Johansen, Groupware: Computer Support for Business Teams, The Free Press, 1988.
- D. Pinelle, C. Gutwin. Groupware walkthrough: adding context to groupware usability evaluation. In *Proceedings of CHI '02*, 455–462. ACM Press, 2002.
- D. Pinelle: Improving Groupware Design for Loosely Coupled Groups, PhD Thesis, University of Saskatchewan, Canada, 2004.

Acknowledgments

Our thanks to Professor Sergio F. Ochoa for his valuable contributions. This work has been partially supported by Fondecyt (Chile) grant No. 1080352.

Developing Mobile Collaborative Applications

Sergio F. Ochoa

Department of Computer Science – Universidad de Chile
sochoa@dcc.uchile.cl
www.dcc.uchile.cl/~sochoa

1. Outline of the Talk

- Introduction to Groupware Systems
- Mobile Groupware Systems
- Groupware Life Cycle
- Design Challenges
- Related Work
- Conclusions and Future Work

2. Collaborative Applications

- Collaborative Applications or Groupware Applications are software that allows a group of persons to collaborate in order to get a common goal.

3. Collaborative Applications

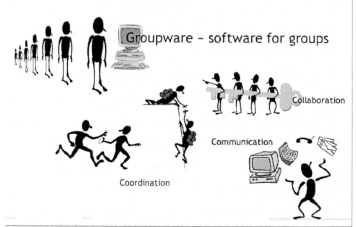

Some examples of collaborative applications:

4. Shared Workspaces

5. Group Calendar

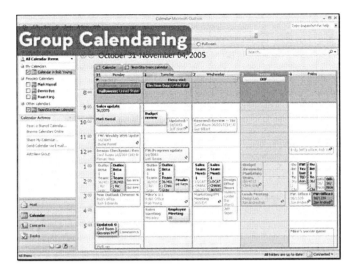

6. Brainstorming

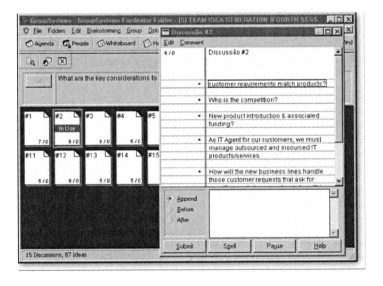

7. Discussion Forums

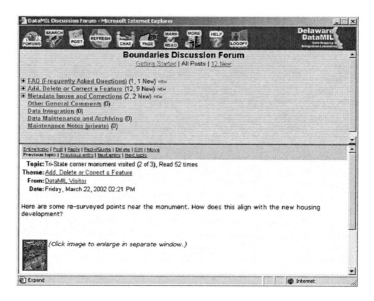

8. Other Applications

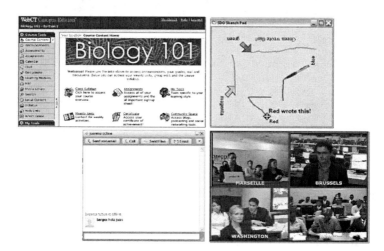

9. Groupware Applications' Features

- Highly demanding for **communication** and **coordination** services
- Client-Server architecture

- Stable communication among nodes

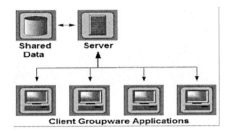

10. Groupware Design Aspects support to oversee the activity

- Users / sessions management
- Management of shared information / objects
- Sync/Async Messaging
- Roles
- Awareness
- Usability, etc.

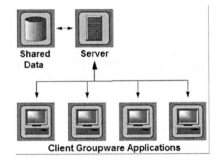

11. Mobile Groupware

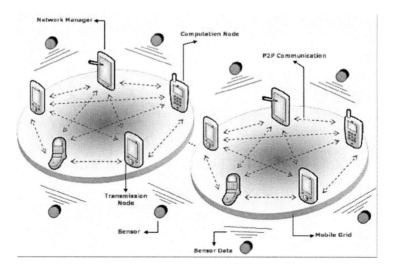

12. Mobile Groupware

- Users / sessions management
- Management of shared information / objects
- Sync/Async Messaging
- Roles
- Awareness
- Usability, etc.

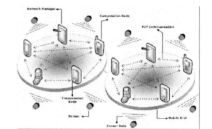

13. … but also…

- Autonomous users
- On-demand collaboration
- Lightweight solutions
- Provide comm. support
- Data synchronization
- Peers detection
- Context-aware
- Interoperability
- Etc.

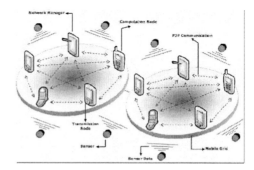

14. Application Scenario: Disaster Relief

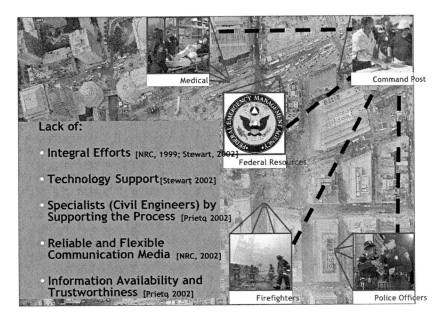

15. Application Scenario: Construction Management

Current situation in construction inspections

Future situation based on MSW usage

16. Application Scenario: Health Care

Informal Communication used to [Tentori & Favela, 2008]:
- Coordinate activities
- Collaborate with colleagues
- Access resources

17. Application Scenario: Health Care

- Informal Communication used to:
    - Coordinate activities.
    - Collaborate with colleagues.
    - Access resources.

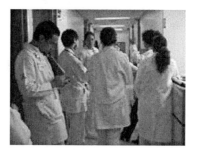

18. Application Scenario: Health Care

- Breakdowns associated to informal collaboration:
- Interruption of interactions.
- Agreements established during the interaction process are not recorded.
- Missed opportunities for collaboration.

19. Again…Mobile Groupware Requirements

- Users / sessions management
- Management of shared information / objects
- Sync/Async Messaging
- Roles
- Awareness
- Usability, etc.

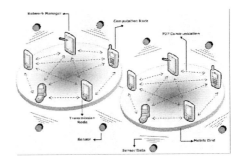

20. Again…Mobile Groupware Requirements

- Autonomous users
- On-demand collaboration
- Lightweight solutions
- Provide comm. support
- Data synchronization
- Peers detection
- Context-aware

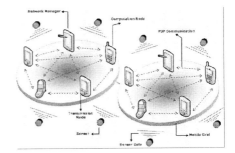

21. Analysis & Design Processes

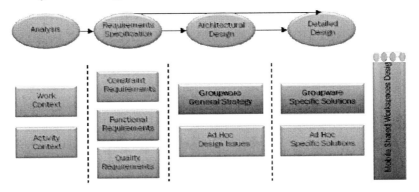

22. Analysis & Design Processes

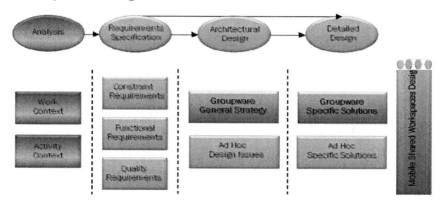

23. Work Scenarios

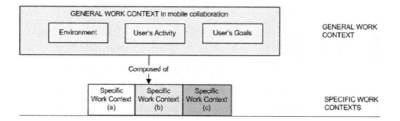

24. Work Scenarios

- What kind of work scenarios will be involved?
- … those with little comfort?
-

25. Work Scenarios

- What kind of work scenarios will be involved?
- ... those with little comfort?
- ... those involving high mobility of people?

26. Work Scenarios

- What kind of work scenarios will be involved?
- ... those with little comfort?
- ... those involving high mobility of people?
- ... those involving a low rate of data input? Networking

27. Work Scenarios: Summary

Desktop PC	Laptop	Tablet PC	PDA/SmartPhone
PDA/SmartPhone	Laptop / Tablet PC		Desktop PC
Desktop PC / Laptop	Tablet PC		PDA/SmartPhone

28. Stabilizing Requirements...

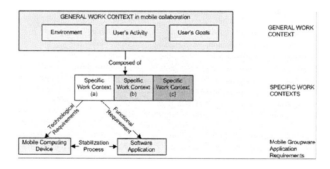

29. Analysis & Design Processes

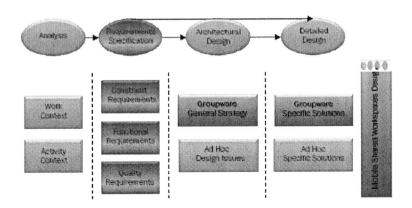

30. Determining the Requirements

31. Again…Stabilize the Req…

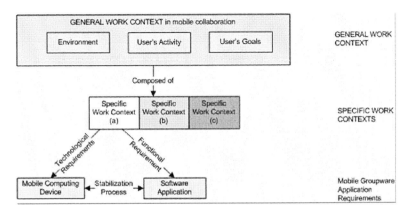

32. Analysis & Design Processes

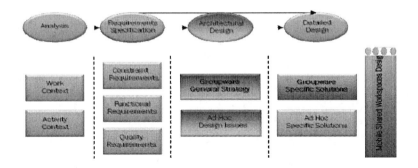

33. Layered Architecture

34. Layered Architecture

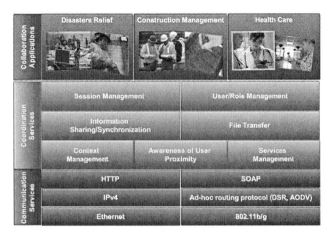

35. Mobile Groupware: Hidden Requirements

36. Which Development Model?

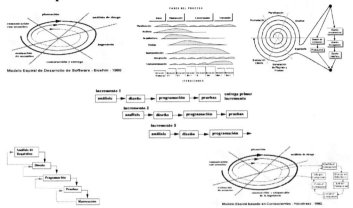

37. Dealing with the Design Challenges

a. Autonomy
b. Communication Support
c. Interoperability
d. Users/Sessions Management
e. Shared Information Support
f. Data Synchronization

38. Users' Autonomy

Fully distribution: allows mobile workers to be autonomous and collaborate among them almost in any scenario.

Goals: independence of centralized resources, deal with wireless networks characteristics (e.g. MANETs).

39. Communication Support: MANET

A MANET is a collection of wireless nodes that can be dynamically self-organized into an arbitrary and temporary topology to form a network without necessarily using any pre-existing infrastructure.

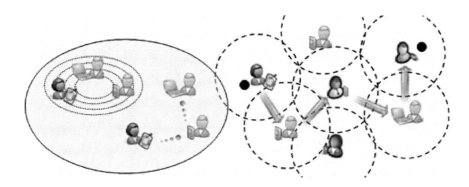

40. Communication Support: MANET

A MANET is a collection of wireless nodes that can be dynamically self-organized into an arbitrary and temporary topology to form a network without necessarily using any pre-existing infrastructure.

At the same time we have:
- wireless communication
- nodes mobility
- routing
- infrastructureless network

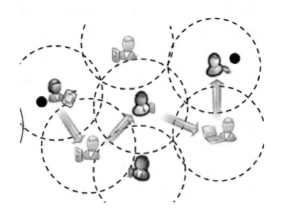

41. Interoperability: Web Services + XML

- Service-Oriented Computing is a paradigm which emphasizes highly specialized, modular and platform agnostic code facilitating interoperability of systems.

- WS and XML are standard technologies.
- These technologies allow to implement privacy mechanisms.
- XML format allows low cost data synchronization.

42. Distributed Sessions Management

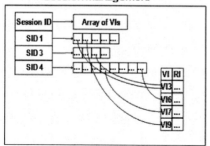

Basic structure of MANET session management

Solution
- Fully distributed
- User's Virtual Identity (VI)
- User's Real Identity (RI)
- User session – an array of virtual identities

43. Distributed Sessions Management

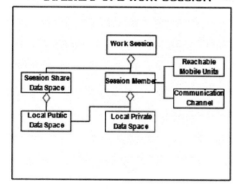

Structure of a work session

Solution
- Work session
- Components
- Session shared data space - public data spaces of active mobile unit
- Session members - a set of mobile units able to interact
- Private data space
- Public data space
- List of reachable mobile units
- Communication channel

44. Mobile Users List

Solution
- The list of reachable mobile units is a data grid.
- Structure of a record:

Mobile Unit ID (Real ID)	Virtual (User) ID	User Role	User Visibility	List of Neighbors

Updating Operations
- Peers discovery.
- Synchronization of users lists.

45. Updating Operations

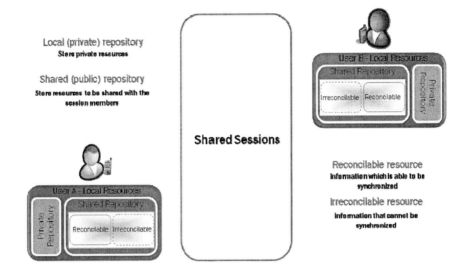

46. Data Sharing & Synchronization

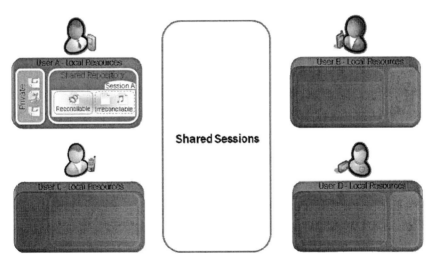

47. Data Sharing & Synchronization

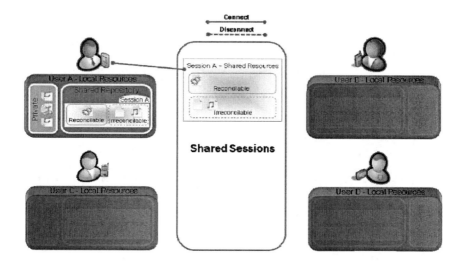

48. Data Sharing & Synchronization

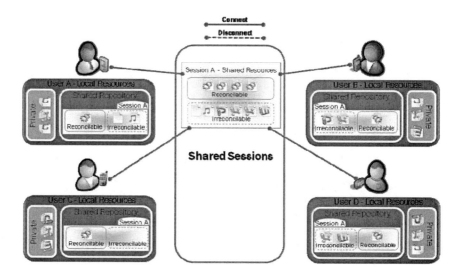

49. Data Sharing & Synchronization

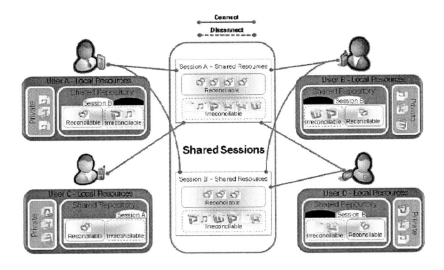

50. Data Synchronization

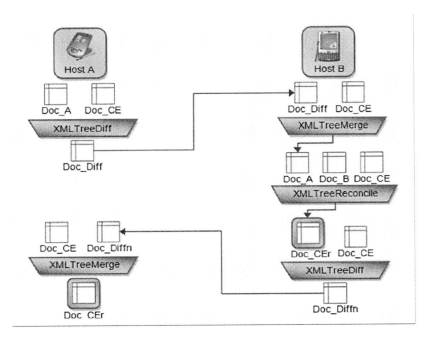

51. Guidelines Summary [Neyem et al, 2008]

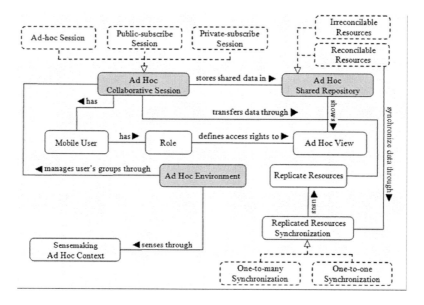

52. Guidelines Summary [Neyem et al, 2007, 2008]

Requirement \ Guideline	Ad-hoc communication	Asynchronous communication	Context-aware solution	Data synchronization	Distributed sessions	Explicit work sessions	Fully distributed	Layered	Platform neutral	Private-subscribe	Reconcilable information	Selection of appropriate mobile device	SOA	Private repository	Shared repository
Discretionary collaboration	▪	▪							▪				▪	▪	▪
On-demand information sharing	▪	▪											▪	▪	▪
On-demand information synchronization	▪	▪		▪							▪				
Session/Users management	▪	▪		▪	▪	▪					▪		▪		▪
Roles support				▪						▪	▪				
Environment support	▪	▪		▪				▪					▪		
Local and shared repositories				▪							▪			▪	▪
Synchronous/asynch interaction support	▪	▪												▪	▪
Autonomous														▪	▪
Context-aware	▪	▪													▪
Easy to deploy			▪									▪			
Self-configurable			▪												
Interoperable				▪					▪		▪		▪	▪	▪
Low coordination cost	▪	▪					▪						▪		
Lightweight													▪		
Asynchronous communication	▪	▪													

53. Related Work

Most of the current initiatives do not consider:
- Ad-hoc communication.
- Interoperability.
- Low Cost Data Synchronization.

54. Related Work

Middleware for Mobile Collaboration	LaCOLLA	JXTA
	iClouds	Nokia Framework
	YCab / YCab.Net	PASIR

Middleware for Peer-to-Peer Data Sharing	LIME	FT-Linda
	JavaSpaces	T-Spaces
	Grace	XMiddle

55. Conclusions

- We are proposing is to use fully decentralized solutions.
- It allows mobile devices to act as autonomous units.
- It allows to share data and synchronize XML documents.
- It allows some interoperability exposing and consuming Web services.
- Of course, these are just the first steps towards stable and useful solutions.

56. Conclusions

- It is a New Area that Requires Exploration.
- The Technological Solutions for mobile collaboration is still Under Development.
- There are Social and Organizational Issues that we are not considering.
- Work on Systems' Usability is Required.
- The Work Context is very Relevant, but does not Easy to Address.
- It could have an important impact in several work areas….

57. Future Work

We have to study:
- Human Behavior.
- User's Interfaces.
- Awareness Mechanisms.
- Test the solutions in real scenarios, and learn of that.
- Evolve towards more reliable solutions.

58. References

[Comfort, 2001] Comfort, L. (2001). "Coordination in Complex Systems: Increasing Efficiency in Disaster Mitigation and Response," *Annual Meeting of the American Political Science Association*, San Francisco, August-September, 2001.

[DHS, 2004] Department of Homeland Security. Research and Technology. Feb. 2004. http://www.dhs.gov/dhspublic/theme_home5.jsp

[Fema, 1999] Federal Emergency Management Agency. (1999). "Federal Response Plan". 9130.1-PL.

[GDIN, 2003] Global Disaster Information Network (2003). URL: www.gdin.org.

[Godschalk, 2003] Godschalk, D. (2003) "Urban Hazard Mitigation: Creating Resilient Cities", *ASCE Natural Hazards Review*. pp. 136-146.

[Guerrero et al, 2006] Guerrero, L., Ochoa, S., Pino, J., Collazos, C. (2006). "Selecting Devices to Support Mobile Collaboration". Group Decision and Negotiation. Springer Netherlands. 15(3), pp. 243-271.

[IFRC, 2003] International Federation of Red Cross and Red Crescent Societies, IFRC. (2003). *World Disasters Report* 2003: Focus on Ethics in Aid.

[Lee, 2002] Lee, R. & Murphy, J. (2002). "PSWN Program Continues to Provide Direct Assistance to States Working to Improve Public Safety Communications". *Homeland Defense Journal*. Vol. 1, Num. 22.

[Mileti, 1999] Mileti, D. (1999). "Disasters by Design: A Reassessment of Natural Hazards in United States". Joseph Henry Press. Washington D.C.

[Neyem et al, 2007] Neyem, A., Ochoa, S., Pino, J. (2007). "Designing Mobile Shared Workspaces for Loosely Coupled Workgroups". Proceedings of the 13th International Workshop on Groupware (CRIWG 2007). Lecture Notes in Computer Science, Vol. 4715, Springer. Bariloche, Argentina. Sept. 16-20, 2007. Pp. 173-190.

[Neyem et al, 2008] Neyem, A., Ochoa, S., Pino, J. (2008). "Coordination Patterns to Support Mobile Collaboration" Proceedings of the 14th International Workshop on Groupware (CRIWG 2008). Omaha, NE, USA. Sept. 15-18, 2008. Lecture Notes in Computer Sciences, (In press).

[Tentori & Favela, 2008] Tentori, M., Favela, J. (2008). "Activity-Aware Computing for Healthcare". IEEE Pervasive Computing, 7(2), pp. 51-57.

[Turk, 2006] Turk, Z. (2006). "Construction Informatics: Definition and Ontology". Advanced Engineering Informatics, 20, pp. 187–199.

[Zurita & Nussbaum, 2006] Zurita, G.; Nussbaum, M. (2006). "An Ad-Hoc Wireless Network Architecture for Face-to-Face Mobile Collaborative Applications". Lecture Notes in Computer Sciences, LNCS 3894, pp. 42-55.

Gesture-Based Interaction

Nelson Baloian [1], Felipe Baytelman [1], Dusan Juretic[1], Gustavo Zurita [2],

[1] Department of Computer Science – Universidad de Chile
nbaloian@dcc.uchile.cl
[2] Department of Management and Information Systems – Universidad de Chile
gnzurita@fen.uchile.cl

ABSTRACT: Nowadays, there is a growing number in different types of computers, with different sizes, shapes and especially input mechanisms. Newer models seem to take a wider distance from the already traditional keyboard and mouse interaction paradigm and getting closer to a more Gesture-based interaction. We see that in electronic whiteboards, tablet-PCs, PDAs and now in intelligent mobile phones. This work presents the works developed since 1994 in which the gesture-based interaction paradigm has been used. First, we present some works using electronic whiteboards in the frame of a project aimed at introducing computer supported collaborative learning in the classroom. This project considered the teacher having an electronic whiteboard (e-beard) to present multimedia learning material and to control the distribution, collaborative use and collection of it among the students. Here, we witnessed the fist problems by not considering this interaction situation as a new one. This moved us to consider a more gesture-based oriented interaction model when developing the interfaces of the software which had to be designed for the teacher to be used on the e-board. In the second part of this work we present some experiences in applying the gesture-based approach to mobile computing. Interestingly, here we discovered that many principles we applied to the design of e-board interfaces can be also applied to the small-size touch screen. In this section, we also present some applications that combine the use of small PDAs with tablet-PCs. We present applications mainly (but not only) applied to the collaborative learning scenario. We also present a platform supporting the development of mobile, peer-to-peer applications in two different aspects. The first one is about the communication in a full peer-to-peer platform. It supports the mutual discovering of the various partners willing to take part in the working (learning) session, and the exchange of information, as well as the recovering of the possible communication failures. The communication platform also includes support for exchanging information using the IRDA feature most mobile devices include. The second aspect is the development of a core library for gesture recognition. This library can be extended to include the recognition of gestures which are particular for a certain application. This library was developed on top of the module used for recognizing gestures for applications implemented to be used with the e-board. This gave us the idea that it would be interesting to have a unified interaction paradigm for different types of computing devices, even if they have different screen sizes. In fact, a unified interaction paradigm will help users to switch from one application to another more swiftly without having to learn new gestures or interaction rules.

1. What is Gesture-based interaction?

- Direct Manipulation
- Gaze-based Interaction
- Movement - Interaction (accelerometer)
- Pen-based interaction

2. Why use Gesture-based Interaction?

- More natural way to interact
- Sometimes is the only way to interact
- Faster way to interact
- Pen-based interaction:
  - Easier to remember
- Especially if it matches "traditional" gestures
  - Faster to perform
  - Fewer errors

3. Motivation

- 1993 IPSI-GMD receives 2 of the first 8 electronic boards released outside XEROX.Palo Alto Labs. (LiveBoard)
- Norbert Streitz PUBLISH: DOLPHIN
  - Ambiente: the disappearing computer
- Ulrich Hoppe COSOFT: Cosoft
  - The Computer-integrated Classroom
- 1995 Move to Duisburg, COLLIDE

4. The Computer-integrated Classroom

The Computer integrated concept development was motivated by bad experiences teaching Java Using off-the-shelf Software. Frequent problems were:

- Having to switch between different application windows
- Too many open windows
- Having to search for specific files to open/send
- Having to type commands

These problems distracted the attention of the students and interrupted the normal flow of the lecture.

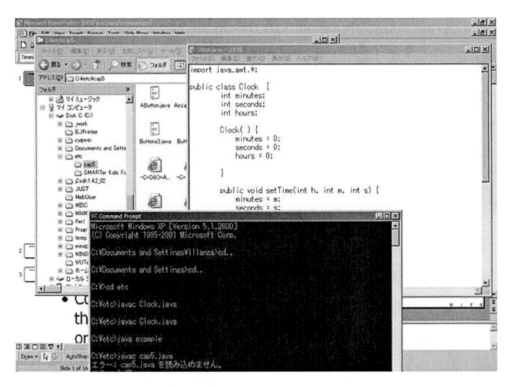

The figure shows how the board's surface may look when teaching java with off-the-shelf software

5. The Computer-integrated Classroom principles

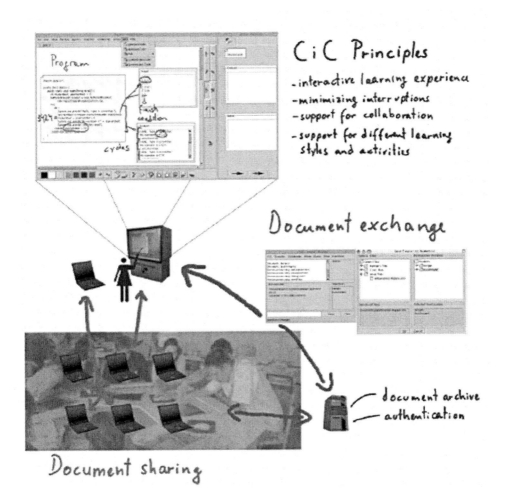

6. The Computer-integrated Architecture

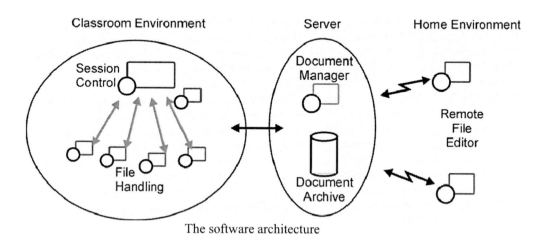

The software architecture

The File system architecture

7. The NIMIS Classroom

Merging of different Media

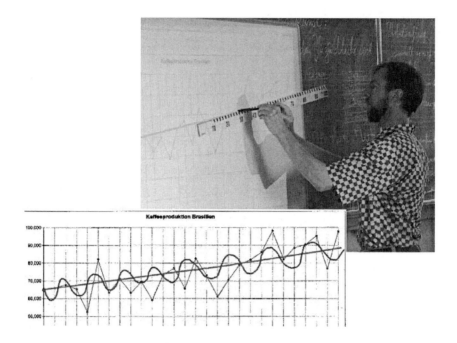

8. Other software systems developed for the Computer-integrated classroom: The Deep Board

- Henning Breuer, post-doc in DCC 2003
- First attempt to develop of gesture-based interfaces for the electronic board
- Developed in Waseda by Felipe Baytelman
- Support for video-based lectures in the GITS between Nishi-Waseda and Honjo

Using DeepBoard for supporting video-based lectures

9. Gesture based interaction support for Mobile computing

- 2005: starting cooperation with Gustavo Zurita
- PhD thesis on mobile support for learning (with Nussbaum, PUC)
- Idea to use gesture-based interaction for mobile devices
  - Natural way to interact with the PDA
  - "Expand" the capabilities of the screen

10. MCSketcher

- The aim of the system was to provide support on the field for small teams of designers
- Interaction based exclusively on gestures, minimizing the number of widgets and the need of a virtual key-board and maximizing the space available for entering content.
- Content organized as a concept map

- No switching between sketching and gesturing
- Full peer-to-peer architecture, no need of central server
- Nomadic computing

Darkened Margin denotes user is in an innersub-node

Design spots show there are other sketching pages attached to this one, which can be explored by following the link shown as a yellow circle. In this case, there are 2 of them

Highlighted Session menu shows the work needs to be saved

"Document three icon"

The "group icon" shows that 2/3 of the users are in this node

The figure above shows a screenshot of the application while a team of landscape designers were using the system to propose changes in a public park. A photograph can be used as background to make the design over it.

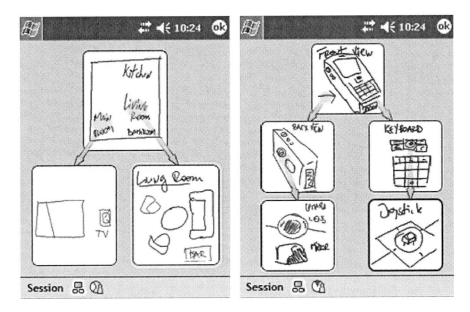

The "tree view" allows an overview of all the spots created and they hierarchy

11. A framework for developing mobile applications

- Gesture recognition module

    - A well designed collection of classes
    - Able to recognize some gestures
    - Can be extended to add more gestures

- A flexible, lightweight communication platform

Motivation: Technology

- Many good desktop applications are programmed on the java platform
- Java applications perform much worse than .NET (with C#) appications on handhelds
- We need a platform which is able to connect both worlds with mnimum effort which takes in account the requirements of both worlds
- Both worlds are object oriented

Requirements

- **Decentralized:**

  - In mobile scenarios, the only available network is the mobile ad-hoc network (MANET) provided by the networking capabilities of the mobile devices.
  - This means, communication and data architecture must follow a peer-to-peer schema avoiding a central server keeping the "master" copy of the data and/or the list of active users.
  - A decentralized peer-to-peer schema adapts itself better to the fact that connectivity between devices is intermittent and the participants list is dynamic, because there is no central server which could leave the session because of a crash or an intermittent communication signal

- In distributed software architecture where communications can fail and applications leave or join the working session dynamically there should be no central server keeping a "master" copy of the shared data and the active users list.
- Therefore, every application must replicate exactly the data others have in order to share a common working environment.
- This means mechanisms must be implemented in order to synchronize the replicated data.
- Since different platforms use different internal object representation schema this is the most convenient way for transmitting an object across different platforms.
- An XML representation of an object contains names and values of the object variables and also some meta-information like the class name, which will be used by the other platform to reconstruct the object.
- There are already "standards" defining the way how an object should be represented by an XML description, we use one of them in our solution.

The Middleware

- Consists of a set of classes implementing an API the programmer can use in order to write distributed applications easily.
- These classes are available in Java and C# and implement the necessary mechanisms for converting objects from their internal representations into an XML representation, transmit them and convert the XML representation into the corresponding internal one.
- Provide and start the necessary services for discovering partners in the ad-hoc network and establish connections among the different applications in order to synchronize shared data.

Discovering partners and establishing connection

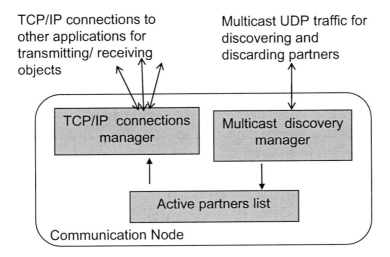

Sharing objects

Data sharing mechanism based on a "shared objects" principle. A shared object is an abstract class which should be extended in order to create an object class whose state will be transmitted to all active participants when the object changes its state, this is when one or more variables change their value.

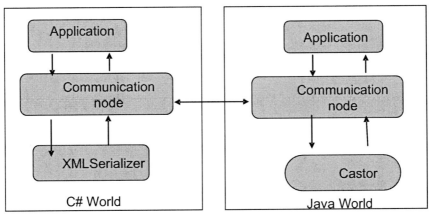

The API

sending/receiving objects	
`public Node(String nodeID, String multicastIP, int multicastPort)`	Creates a Node object which starts the Multicast service for discovering and the TCP/IP server for transferring data.
`public void receiveObject(Object o)`	Used by the communication node in order to receive the objects sent by the partners
`public void sendObject(String partnerID, Object obj)`	Sends an object to a certain partner. If partnerID is null the object will be sent to all partners in the network
`public void sendObject(String[] usrIDList, Object obj)`	Sends an object to a list of users
`public void sendToGroup(String groupName, Object o)`	Sends an object to all partners registered in a specific group
Group management	
`public void join(String groupName)`	Joins the application to a certain group characterized by the group's name
`public void leave(String groupName)`	Detaches the application from the group specified
`public void remoteJoinGroup(String groupName, String partnerID)`	Invokes the join method in a remote application, forcing that application to join a group
`public void remoteLeaveGroup(String groupName, String partnerID)`	Invokes the leave method in a remote application, forcing that application to leave a group

A Platform for Motivating Collaborative Learning Using Participatory Simulation Applications

Gustavo Zurita [1], Nelson Baloian [2], Felipe Baytelman [1], Antonio Farías[1]

Department of Management and Information Systems – Universidad de Chile
{gnzurita, fbaytelm, anfari}@fen.uchile.cl
Department of Computer Science – Universidad de Chile
nbaloian@dcc.uchile.cl

1. Aims, Goals

- Design a conceptual framework

  - For developing applications of participatory simulations
  - Motivating collaborative learning

- Explore the role of handhelds supporting Participatory Simulations

  - Exchange data in ad-hoc mobile environment
  - Each handheld/user acting as an agent in participatory simulation
  - Actively learning agent roles and outcomes

- How good supports participatory simulation the learning process ? Which scenarios are the best.

2. Participatory simulations

- Role-playing activity oriented towards learning complex and dynamic systems

  - Mapping real world problems to simulated context and behaviors
  - Knowledge and patterns emerge from local interactions among users

- Highly effective in large groups

  - Simple to set up and interact with
  - Simple decision process: Analyze information, exchange information, make decisions and see the outcomes

- It allows to relate actions and their consequences

- Highly motivating even in large groups

    o Participation and collaboration increase the understanding of the simulated reality and problem-solving abilities
    o Mobility has positive effects in engagement
    o Can be integrated in a whole classroom
    o Learning by doing.

3. Related Work

- There is a growing interest in participatory simulation

    o Has shown to simplify several concepts about complex systems (e.g. distribution, synchronization and emergent behaviour)

- Different Soft-hardware devices have been used

    o Spreadsheets over PCs, thinking tags, laptops, handhelds

- For different learning subjects

    o Futures trading, virus propagation, problem-solving, financial theory

- We see a contribution opportunity by developing a framework which would ease the development of applications implementing participatory simulation

    o handhelds,
    o ad-hoc wireless networks and
    o pen-based interfaces: Gestures and sketches

4. The Framework should help:

- Networking
    o Discover participants, build connections, synchronize information

- Define and assign roles
    o Which kind of actors and what they will represent

- Define and assign objects
    o Which kind of objects will be exchanged between participants

- Specify rules
  - o How will objects be exchanged (still not very automatic)

- Teacher support
  - o The conductor must ensure all participants play an active role in the simulation

5. HCI Principles considered

- Pen-Based User Interface

  - o offers a more natural and intuitive interface enabling the sharing and exchange so as to improve efficiency

- Gestures

  - o most frequent actions are deleting, selecting and moving
  - o an efficient as a form of interaction
  - o easy to learn, utilize and remember

- Mobility and exchange of objects on the fly

  - o handhelds provide high mobility and portability
  - o way for creating ad-hoc networks through peer-to-peer connections
  - o such a network allows information exchange between
  - o proximity detection is done with infrared sensors (IrDA) combined with Wi-Fi

6. Designing (creating) roles &Items

Example: A Trust building rules learning scenario

a.   b.

7. Designing Items

Example: Diseases, symptoms and treatments

8. Exchanging Items: Proximity+ IrDA

9. Exchanging Items: Proximity+ IrDA

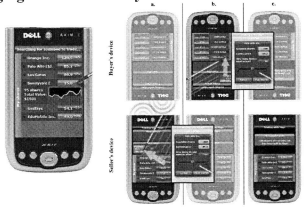

10. Teacher support to oversee the activity

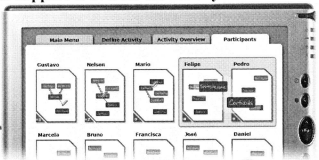

11. Ubiquitous Computing Scenarios

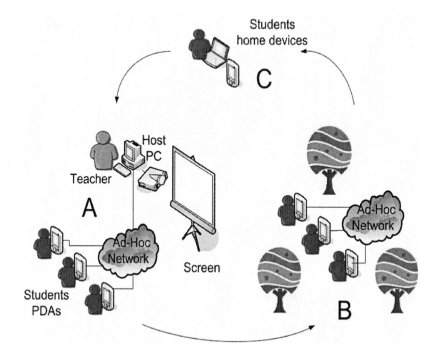

- Integration of computer supported activities in formal and informal scenarios (learning)
- Integration of activities "on-the-field" and inside the office (design, collect data on-site process it)

Supporting rich interaction in the classroom with mobile devices

Gustavo Zurita [1], Nelson Baloian [2], Felipe Baytelman [1],

Department of Management and Information Systems – Universidad de Chile
{gnzurita, fbaytelm}@fen.uchile.cl
Department of Computer Science – Universidad de Chile
nbaloian@dcc.uchile.cl

1. Critical factors promoting social interaction in the classroom

- open problems motivate collaboration by assessment and feedback.
- The teacher provides coaching to the student
- Students exchange artifacts representing their ideas during reflection
- Construction mutual and collaborative understanding (peer-review).

2. Model Of Social Interactions

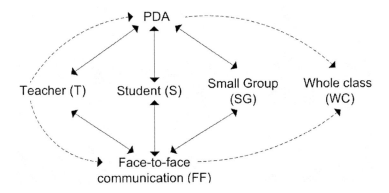

Straight lines represent teacher-student or teacher-small groups interaction; dotted lines represent teacher-whole class interaction from: Lagos, Alarcon, Nussbaum, Capponi: Interaction-Based Design for Mobile Collaborative-Learning Software

3. Design & Implementations Issues

- Pen-Based User Interface

  - o offers a more natural and intuitive interface enabling the sharing and exchange so as to improve efficiency

- Gestures

  - o deleting, selecting, moving, resizing, rotating
  - o efficient as a form of interaction
  - o easy to learn, utilize and remember

- "Shared Objects" platform for developing peer-to-peer applications

  - o Automatic recognition of partners
  - o Transparent distribution of changes of a shared object
  - o Group management
  - o Recovery from errors and temporary disconnections

4. System Architecture

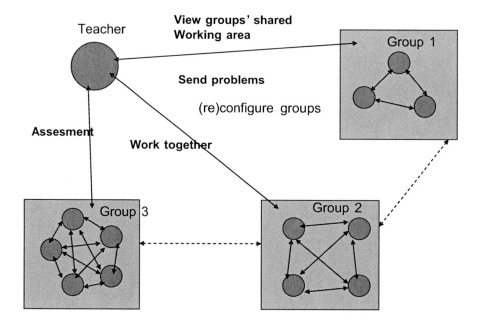

5. Some gestures

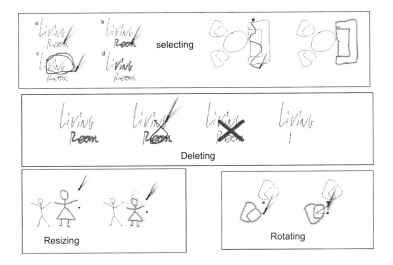

6. Group configuration

Teacher's module, in Group setup mode. Dragging user Nelson to group 2. Users are displayed automatically when discovered. The groups are defined by the teacher.

7. Problem creation

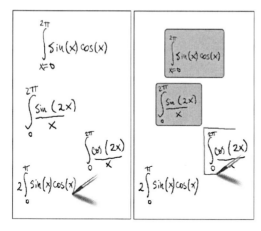

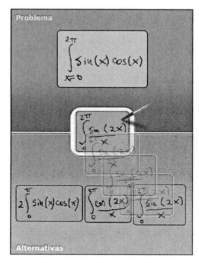

*writing problem definition and
answers and delimiting the elements
by closing them in rectangles*

*For defining problems with
alternatives problem parts are
dragged into the respective areas*

8. Synchronized work

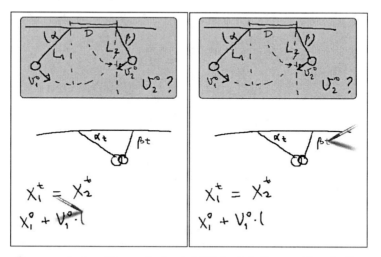

*Screenshots of two students' PDAs jointly working in the
same group work solving different part of a problem*

9. Sending the answers

Right: Button shows the students must chose an answer
Center: Button turns into and agreement indicator.
Left: Once all members agree, button turns into "Submit answer"

10. Assessment

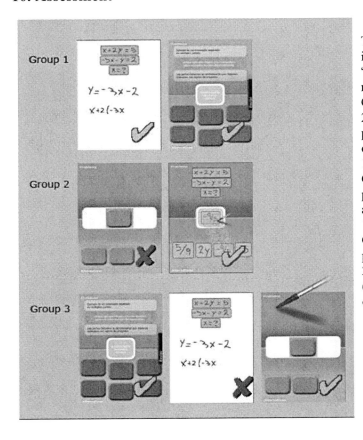

The teacher' view in the "show results" mode.
Group 1: answered 2 problems right (1 open and 1 with alternatives).
Group 2 answered 2 problems with alternatives, 1 wrong 1 right
Group 3 answered 3 problems 2 with alternatives (right) one open (wrong).

11. First tests

- During two weeks, 24 students and two different teachers from a pre-graduate university course

- All type of problems where used including group reconfiguration

- System used twice a week in sessions of one and a half hour each.

- Teacher used a Tabled PC and the students used PDAs.

12. Results

There was a high level of social interaction. Teacher could easily and swiftly perform forming groups, distribute exercises and students' assessment activities Group reconfiguration and inter-group interaction eased by mobile devices.

Results of a survey
- System was easy to learn: 67% "agree", 33% "strongly agreed".
- I felt comfortable using sketches, gesture based interaction 75% "agree" 25% "strongly agree"
- The system supports interaction 66.6% "agree", 16.6% "strongly agreed", 16.6% " neutral"
- All students agreed that system motivates face-to-face interactions with other students and with the teacher due to mobility. Both teachers agreed that the operation of the system was simple. Comments suggested collaborative editing of the sketches was difficult at the beginning

Contributions of this work
- It is a full peer-to-peer architecture (can be used outside the classroom)
- independent of the learning subject (problems and solutions based on free-hand writing and sketching)
- creation of new problems "on the fly" (evolving problems)
- It allows the teacher to overview all the workspaces of the groups at the same time.

MCPresenter: A Collaborative PDA & Table PC Based Tool to Support Pedagogical Practices

Dušan Juretić, Nelson Baloian, Gustavo Zurita

1. General Context

- Using traditional methodology in the classroom the interactivity level is low. For example, using slideware as Power-Point.

- In general, the most skilled students participate in class, while others don't because of shyness or the size of class. [RJ. Anderson et al. 2003]

- With the advance on the technology field, classrooms have started to use devices as Tablet-PCs, PDAs, etc.

- Learning outcome of students is increased if they do activities by themselves rather than just listening ("learning by doing").

- Low interactivity => teacher can't know accurately how well are the students learning...

- ... until a test/quiz has taken place (too late to modify teaching methods).

- Mobile Collaborative Presenter (MCPresenter) is an application to support teachers' work in a classroom, based on the collaborative learning using technological tools.

- MCPresenter allows bidirectional communication between teacher and students, so the teacher can know the learning

- level of his/her students, and students can obtain feedback about their responses.

- Uses a Wi-Fi ad-hoc network.

- This application is under development, is written in C#, under two versions: using .NET Framework 2.0 (for PC/Tablet-PC) and .NET Compact Framework 2.0 (for PDA), on the top of NOMAD.

2. Mobile Collaborative Presenter (2)

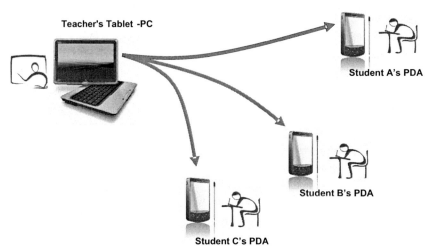

Figure 1: Delivery of class material to students

3. Features

- Implemented features:
    Teacher's area.
    Student note-taking.
    Creation/modication of presentation paths.

4. Teacher's area

- Teacher owns a "special" area where he/she can write freely.
- Students have to ask permission through the system to write there.
- Teacher can give/remove permission to write on the teacher's area.

5. Teacher's area (2)

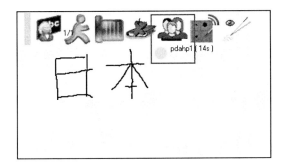

Figure 2:   Teacher viewing students' permissions to write

6. Note-taking and note-sharing

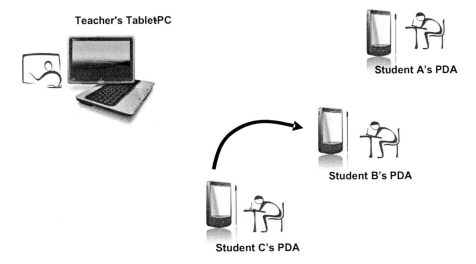

Figure 3:   Sharing notes to a specific person

7. Note-taking and note-sharing

Figure 4: Sharing and seeing shared notes

= Collaborative Note Taking

8. Defining presentation path

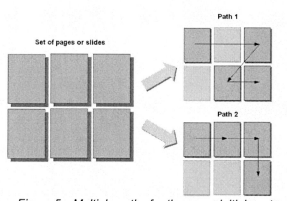

Figure 5: Multiple paths for the same initial content

If the teacher wants to...
- ... view students' work, select some to be shown to the class.
- ... modify the order of the material to be shown.
- ... save path for future classes.

=> Presentation Path

9. Defining presentation path (3)

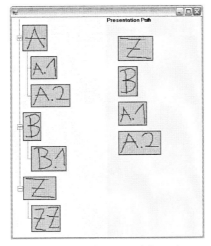

Figure 6:   *"Defining path" mode*

10. Future work

- Complete the presentation path feature (multiple paths, navigate nodes according to path)
- Simplify access to options.
- Load slideshows from PPT les (and decide what to load from them).

11. Future work (2)

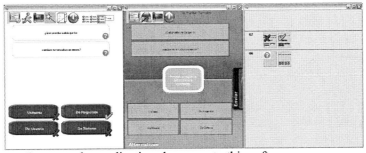

- Integrate the application shown up to this software.
- Show additional information about quiz results (for the teacher)
  - Overlap students' responses of an open problem.
  - Show a graph showing students' choice in multiple choice questions.

Algorithm Visualization Using Concept Keyboards

N. Baloian, H. Breuer, W. Luther,
Ph. Kraft, Chr. Middleton, Th. Putzer

Departamento de Ciencias de la Computación, Universidad de Chile, Santiago
Institute of Informatics and Cognitive Sciences, University of Duisburg-Essen

nbaloian@dcc.uchile.cl, luther@inf.uni-due.de

Abstract. The focus of our talk is an interactive approach to studying algorithms and data structures with different kinds of concept keyboards. It is a significant result that people prefer a step-by-step keyboard emulated on the screen to study an unknown algorithm and a hardware concept keyboard to test the functionality of the algorithms specific methods. As a consequence, modern information and learning systems for algorithm animation are enhanced in such a way that control and interaction take place through appropriate interfaces designed and semi-automatically generated for this special purpose. In this overview, we provide some examples and report on a thorough evaluation to show the relevance of this new approach.

Keywords: Concept keyboards, algorithm animation, algorithm visualization, XML-based interface description

1 Content

Overview
- Proposal and requirements
- Goals
- Concept keyboard software CONKAV
- Software characteristics
- Multimodal features
- Configuration and application
- Algorithms implemented
- Tests
- Evaluation
- Results and analysis
- Conclusion and future work

2 Introduction

Project context
- Importance of Algorithm Visualization (AV)
  - AV helps instructors to explain
  - AV helps students to understand
- Algorithm animation and visualization tools present
  - Relevant data
  - Current states
  - Sequences of appropriate visualizations of objects
  - Tools to generate animation and visualization

3 Visualization for the Mind's Eye

- Use new concepts for enhancing standard visual interfaces with aural or haptic components
- Convey logical structures or data types to human minds [1]
- Optional dual or multiple interfaces enhance the human-machine interaction and support sensory-disabled people
- Separate control elements and visual objects by mean of an adequate concept keyboard
- Process data structures and geometric models
  - Focus on important information after a suitable abstraction process.

4   Algorithm Comprehension

- Conventional AV systems
  - Show a graphical representation
  - Allow to present step-by-step
- Research on pedagogical benefit of AV
  - To obtain better understanding
    - Students should be more involved
    - Allow students to do things
  - "What learners do, not what they see, may have the greatest impact on learning" (Hundhausen)

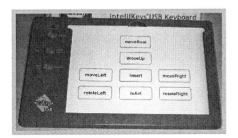

5 Algorithm Manipulation

- Allow enhanced control
- Create relationship operation / user interface
  - Use Concept Keyboards (CK)

6 Proposal

- Students can achieve better comprehension by navigating through the data structure by
  - Explorating
  - Experimenting
- CONKAV helps to **understand how the algorithm works.**
  "I hear and I forget, I see and I remember,
  I do and I understand" (Konfuzius)

7 Concepts

- Define a level of detail the user is supposed to learn about the algorithm
- Develop for each algorithm a suitable CK by redefining keys or creating special button schemes
- Develop a spatial arrangement of the keyboard
- Define appropriate output channels
- Let the user choose which step should be executed
- Include an efficiency control that gives the user optional feedback on the quality of his solution.

8 Requirements

- Use existing implementations
- Describe functionalities and use methods
- Semiautomatic generation of the keyboard
- Customization
- Algorithm visualization
- Interactive manipulation of data structures
- Undo and redo
- Initial data files (scenarios).

9 Goals

- Main Goal
  - Determine if better comprehension is achieved by allowing the student to explore and manipulate the data structure
- Specific Goals
  - Determine requirements and characteristics
  - Design and develop configuration software
  - Design and develop visualization/interaction software
  - Design and execute tests
  - Prepare campus wide use of CONKAV.

10 Software Characteristics

We designed an architecture providing a number of components. The methods implementing the algorithm are described in the AGIXML file as are the data structures related to the algorithm, for instance, an array of numbers (to be sorted) and a pointer defining the focus on a node of the tree or an integer value as the total number of nodes. The first component called algorithm implementation (AGI) corresponds to the principal class implementing the algorithm, which is loaded on the fly by reading its classname and location. The Reflection API allows a Java application to acquire the definitions of classes and operate on them at runtime without having them available at compile time: Inspect and modify. The library supports the instantiation of a class from its text name, acquiring information on a class and calling methods on an object. Once an instance of a class has been created, the controller can access meaningful methods that execute the algorithm action-by-action with a certain granularity for manipulating the data structure. The IOXML file specifies the CK or alternative input devices used, a data input frame and one or several output devices.

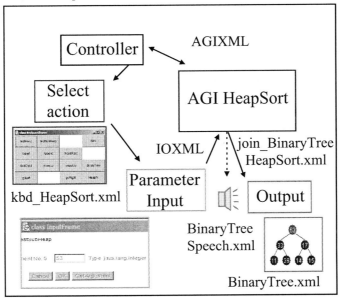

Multimodal system design for the HeapSort algorithm

11 Tactile overlay for CK and the HeapSort algorithm (Th. Putzer)

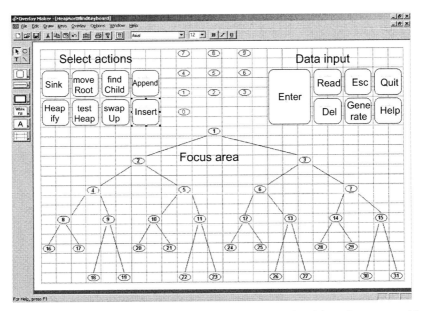

Multimodal features were developed so that people with visual impairments could use the software [2].

12 Concept Keyboard Software (Chr. Middleton)

During the Configuration stage, the user acts as a designer, selecting the methods within the class of the implemented algorithm that will be necessary to execute it with a certain granularity.

During the second stage of the configuration, the user designs the layout of the CK. This process consists in determining the position of each of the keys on the keyboard. The user drags and drops the buttons representing the "actions" of the algorithm to positions on the keyboard. The keyboard is initially represented by a 5×5 grid, which can be extended to any M×N grid or be freely generated in a forthcoming version. The idea is that the designer groups the buttons of closely related actions together [2].

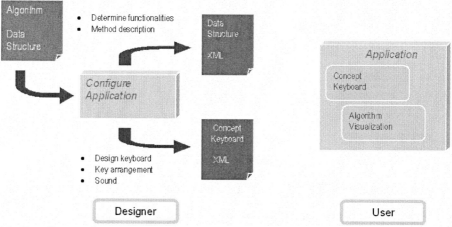

13 Configuration

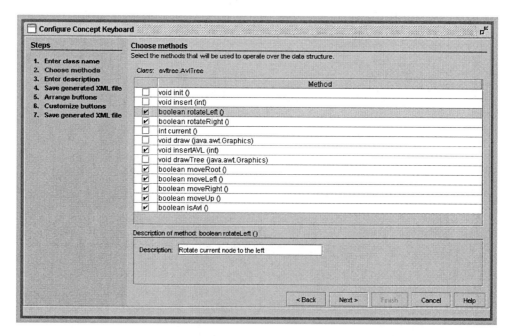

14 Application

During the visualization of the algorithm, the user can load different algorithms and their corresponding keyboards. Each of the algorithms has different CKs, depending on the actions the user can execute. Furthermore, the system allows the user to load different layouts of the same CK for each of the algorithms. It is also possible to load a step-by-step keyboard and the startup file (if the algorithm allows it) to change the initial values of the graph, tree, etc., in order for the student to have different problems to solve.

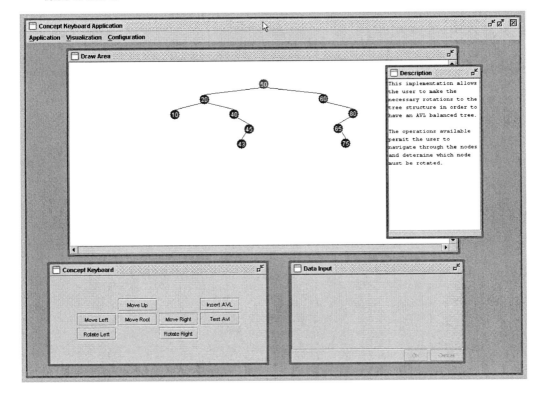

15 Algorithms implemented

Trees algorithms:
- AVL trees

Graph algorithms:
- Dijkstra (shortest path)
- Kruskal (minimum spanning tree)
- Prim (minimum spanning tree)

Sorting algorithms:
- QuickSort
- HeapSort

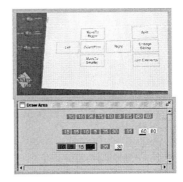

16 Evaluation (H. Breuer)

First, we conducted two qualitative evaluations as they are usually made before setting up an extensive quantitative evaluation. The intent of such formative or exploratory evaluation is not to gather representative data about the quality of the system under evaluation but to generate ideas concerning how to improve the system and what aspects to focus on within a quantitative evaluation. In the following we list some of the subjective responses we received from these initial evaluations followed by the hypotheses we derived from them that would direct the quantitative evaluations.

"Students can view the results of the modifications made so that they can test whether they have understood the algorithm. This allows them to study specific cases about which they have doubts, instead of using only the pre-made examples available on the web." This utterance, among others, was summarized in Hypothesis 1: The use of the CK fosters the comprehension of algorithms.

"The user can choose which actions to take; it is not just step-by-step solving." "It provides greater interaction [with users], and allows them to learn from their mistakes." These statements let us formulate Hypothesis 2: The enhanced exploration modes of the CK approach facilitate the learning process.

After evaluating the questionnaires, we concluded that all but one student preferred to work with the Visualization-with-Keyboard system. So we decided to examine Hypothesis 3: Visualizing algorithms with the CK is more motivating than working with the step-by-step keyboard.

Hypothesis 4: The step-by-step keyboard is appropriate for beginners was chosen because we received comments such as this: "When learning a new algorithm (as I am), I prefer the [web] visualization (with the step-by-step interface)."

Participants were asked to evaluate both the CK and the step-by-step interface. Most items took the form of assertions, and students declared their level of agreement on a five-point Likert scale ranging from 1 to 5.

The students that used the Visualization-with-Keyboard system before solving the exercises answered 95% of questions correctly. In contrast, those that used the Web Visualization and the step interface before solving the exercises answered 90% correctly. This result led us to Hypothesis 5: The group using the CK performs better in the posttest than the group using the step keyboard [4]. As a last step in the two-hour evaluation process, participants were instructed to configure their own keyboard.

17 Execution of Tests

- Divided into several stages
  Santiago, Chile (December 2003)
  Duisburg, Germany (January, November 2004 - January 2005)
- Santiago 2003
  Compare CONKAV to Web visualization
  Determine improvements required
- Duisburg 2004
  Compare Concept Keyboards to adapted step-by-step interfaces
- Duisburg 2004 - 2005
  Compare hardware CK to Wacom tablet, design keyboards

18 Tests in Duisburg (Ph. Kraft)

November 2004 – January 2005
- 100 students from a second year course on A&DS
- Work with different implementations of CKs
- Answer to a questionnaire of seventy-seven items consisting of open and closed questions
- Solve a set of exercises (pre- and post-performance test)
- Configure different concept keyboards for the QuickSort, AVL-trees and Dijkstra algorithm.

19 Results

- The CK software displaying three windows on the screen was generally approved (M 4, R 1-5)
- Customize labels of the keys are welcome (M 4, R 1-5)
- Participants suggested increasing the number of algorithms presented and adding an extended explanation of the algorithm being studied
- Students like receiving a verbal description of the characteristics and functioning of the algorithm
- The tool provides better insight into the proposed algorithms and data structures (Median 4, Range 1-5)
- Twenty-five persons preferred the hardware CK, nineteen the Wacom tablet and ten the virtual CK, while six were undecided
- Compared to the step keyboard, the CK was favored by thirty-seven students
- Students with lower knowledge preferred the step-by-step keyboard
- Concept keyboard users have significantly better results (Mann-Whitney test).
  - Concept keyboard users can explain the algorithms in a better way than step by step – keyboard users
- A majority of users likes to try out several ways to solve the problems and the separation of input and output
- Students perform better with the hardware concept keyboard than the Wacom tablet
  - Shorter answering time
  - Less steps
  - More fun.

Finally, all participants were asked which interface they would prefer for learning algorithms. The CK was preferred by 62% of the participants, whereas 38% preferred the step interface.

20 Conclusion and Further Work

We have presented a software architecture for a new controller design that allows users to select a class library on the fly, to instantiate objects and to call their objects. Input and output interfaces are configured through XML files.

Experimenting with algorithms through these new interaction and control elements lead to a better understanding.

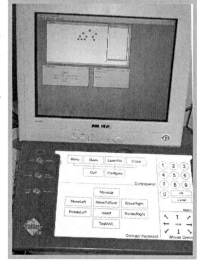

Further improvements:

- Implement a database of students' results
- Add new perception channels
- Different sounds
- Tactile feedback (use layouts)
- Version for blind people
- Use a different granularity of the methods
- Add new algorithms

21 Outlook

A new cooperative CK software intended to learn interactive cryptographic protocols was recently tried out in a standard lecture on cryptography at the University of Duisburg-Essen with 40 students. An initial evaluation confirmed the results attained in our previous evaluations and proved the usability of the CK software when applied to a variety of algorithms.

References

1. Baloian, N. and Luther, W.: Visualization for the Mind's Eye. Workshop on Software Visualization, Dagstuhl, 20.-25.05.2001, Software Visualization, State-of-the-Art Survey LNCS 2269, (St. Diehl ed.) Springer, 2002, pp. 354-367.
2. Baloian, N., Breuer, H., Luther, W., Middleton, Chr.: Algorithm visualization using concept key-boards. Proc. ACM SoftVis'05, St. Louis, Missouri, May 14-15, 2005, pp. 7-16.
3. Baloian, N., Luther, W., Putzer, Th.: Algorithm Explanation Using Multimodal Interfaces. IEEE Press CS SCCC 2005, Valdivia, November 7-12, 2005, pp. 21-29.
4. Baloian, N., Breuer, H., Luther, W.: Concept keyboards in the animation of standard algorithms, Journal of Visual Language and Computing, 19, 2008, pp. 652–674. doi: 10.1016/j.jvlc.2007.07.00510.

Cooperative Visualization of Cryptographic Protocols Using Concept Keyboards

Nelson Baloian, Wolfram Luther

Departamento de Ciencias de la Computación, Universidad de Chile, Santiago
Institute of Informatics and Cognitive Sciences, University of Duisburg-Essen
nbaloian@dcc.uchile.cl, luther@inf.uni-due.de

Abstract. This talk mainly presents the software called CoBo (Cooperative exploring and visualizing cryptographic protocols using concept keyboards), which applies the principle of the "concept keyboard" to implement a system that supports the learning of cryptographic protocols. An initial evaluation confirmed that the use of the CK fosters comprehension of the algorithms, facilitates the learning process and stimulates the learners' activity.

Keywords: Concept keyboards, cryptographic protocols, cooperative exploration and visualization

1 Introduction

The aim of the present work was to develop a collaborative software tool called CoBo. Based on the hypothesis that collaborative learning activities motivates the learner and lead to better results [1], this tool would help visualize various cryptographic communication protocols using concept keyboards to interactively control execution in a collaborative way. The result was a multi-user software tool that allows users working at different computers or PDAs to replay the various cryptographic protocols and collaboratively explore, analyze and share the results of their actions.
In a recent paper Cattaneo et al. [2] presented GRACE. (graphical representation and animation for cryptography education), a Java-based educational tool that can be used to help in teaching and understanding of cryptographic protocols. GRACE implements and visualizes well-known protocols using several cryptographic primitives and cryptosystems. The visualization is controlled by a step interface, and the tool does not provide cooperative work.

2 Four Protocols Implemented

The CoBo system was conceived in order to implement a Computer Supported Collaborative Learning System based on the visualization and simulation of cryptographic protocols using concept keyboards.

- Wide Mouth Frog (WMF)
- Feige-Fiat-Shamir (FFS)

- Needham-Schroeder
- Kerberos V

3 System Architecture - XML-Files

The architecture of CoBo was conceived in order to define most of the functionalities of a new protocol with XML files.

XML files:
- Scenario file
- Facts basis file
- Algorithm file

Standards used:
- JDM
- SwixML
- PetML
- XSLT

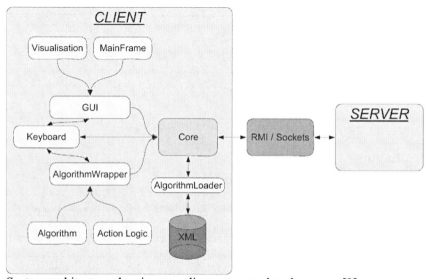

System architecture showing one client connected to the server [3]

4 Scenario File

- Complete name of the so-called Wrapper-class of the protocol
- Path for the XML file describing the "facts" of the protocol
- Short description of the protocol
- Listing of the actors (roles) involved in the protocol.

5 Facts Basis and Protocol Description File

- Keyboard Definition: This part describes the graphical layout of the concept keyboard
- Method Mapping: Methods of the protocol programmed in the AlgorithmWrapper class are assigned to a key
- Protocol description: Contains an HTML description of the protocol
- Action logic: XML description of the protocol.

6 Action Logic Generation for Program Description

- Options to implement the action logic
- Stack machine
- Finite state automaton
- Petri-StateMachine (PSM)
- self-executable and text-executable transitions

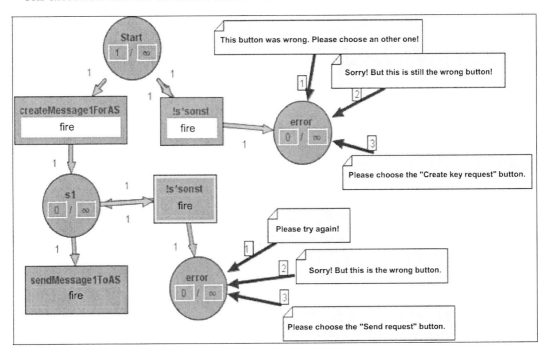

Part of the Petri net for the Needham-Schroeder protocol [3]

7 Configuration Dialogue for an Individualized Keyboard

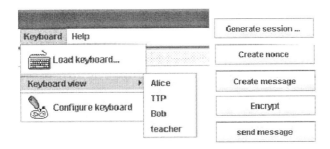

WMF-protocol – teacher's view (left, right)

8 Visualization

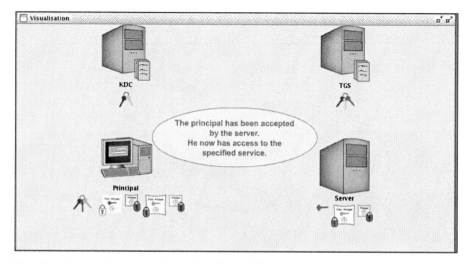

Needham Schroeder protocol (A. Kováčová, [4])

The various actors participating in the simulated cryptographic protocol are shown, as are the elements generated during its performance (messages, keys, encrypted messages, etc.) providing a clear visualization of the "knowledge" accumulated at this state of the protocol. This window also shows the "animation" of the protocol.

Actors

| Alice A | Bob B | Authenticity center AS | Intruder I |

9 Needham-Schroeder Protocol

Symmetric encryption standard

t_A: Time stamp; **CK** : Communication Key; **KA, KB**: Prechared Keys

1	$A \rightarrow AS(A,B,t_A)$
2	$AS \rightarrow A(t_A,B,CK(CK,A)^{KB})^{KA}$
3	$A \rightarrow B(CK,A)^{KB}$
4	$B \rightarrow A(t_B)^{CK}$
5	$A \rightarrow B(t_B-1)^{CK}$

I(A) - Intruder imitates Alice (I knows CK)

3*	$I(A) \rightarrow B(CK,A)^{KB}$ *replay attack*
4*	$B \rightarrow I(A)(t_B)^{CK}$
5*	$I(A) \rightarrow B(t_B-1)^{CK}$

Man-in-the-middle-attack in the asymmetric version and corrected version

1	$A \rightarrow AS(A,I)$	$A \rightarrow AS(A,B)$
2	$AS \rightarrow A(PKI,I)^{SKAS}$	$AS \rightarrow A(PKB,B)^{SKAS}$
3	$A \rightarrow I(t_A,A)^{PKI}$	$A \rightarrow B(t_A,A)^{PKB}$
4	$I(A) \rightarrow B(t_A,A)^{PKB}$	
5	$B \rightarrow AS(B,A)$	$B \rightarrow AS(B,A)$
6	$AS \rightarrow B(PKA,A)^{SKAS}$	$AS \rightarrow B(PKA,A)^{SKAS}$
7	$B \rightarrow I(A)(t_A,t_B)^{PKA}$	$B \rightarrow A(t_A,t_B)^{PKA}$
8	$I \rightarrow A(t_A,t_B)^{PKA}$	$B \rightarrow I(A)(t_A,t_B,\mathbf{B})^{PKA}$
9	$A \rightarrow I(t_B)^{PKI}$	
10	$I(A) \rightarrow B(t_B)^{PKB}$	$A \rightarrow B(t_B)^{PKB}$

CoBo Graphical User Interface

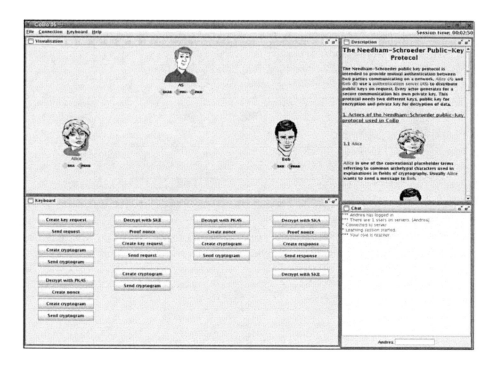

10 Evaluation of the CoBo-System [5]

- Group of forty computer science students (aged 20 to 35, average age 24.28)
- Randomly chosen from a third-year course on cryptography and network security at the University of Duisburg-Essen
- Majority estimated their own knowledge as low (WMF 50%, FFS 57.5%, Kerberos 52.5%).

Two scenarios

- Single perspective
  - Ten hypotheses
  - 45 statements
- Cooperative work - group evaluation
  - 13 groups
  - Seven further hypotheses
  - 18 items

Concept keyboard facilitates understanding

- The use of the CK supports the learning the entire extent of cryptographic protocols. (2, L5, Median 4, Range 1–5), Mean=4.01, Variance=0.89
- The free exploration and the possibility of taking erroneous actions foster the learning process. (4, B2, 75%)
- The CK makes the learning process more efficient. (4, B2, 97.5%)
- Interacting with the CK stimulates the motivation. (3, L5, Median 4, Range 1–4, 74%), Mean=3.93, Variance=0.97.

Keyboard design

- The keyboard is well structured. (1, B2, 95%)
- The error messages are comprehensible. (4, B2, 82.5%)
- The help desk contains all necessary information to support the user during the learning process. (5, B2, 85%)
- Concise error information and explanations in response to specific questions enable the user to learn the protocols. (2, L5, Median 4, Range 1–5, 70%) Mean=3.925, Variance=0.938
- The screen design assists the understanding of the protocol step-by-step. (17, B2, 94%)
- The visualization does not create obstacles for the interaction. (3, B2, 92.5%).

Group evaluation

- The group-based learning provides only a partial view of the protocols to the individual actor. (1, L5, Median 3, Range 1–5, 48%), Mean=2.79, Variance=1.51
- It is possible to allocate the actions to the actors that triggered them by interpreting the visual output. (2, B2, 85%)
- The customized and individualized keyboard promotes deeper understanding of the protocol. (4, B2, 82.5%)
- The use of individualized concept keyboards in the cooperative scenario helps to learn the protocols in an efficient way (3, B2, 86%)
- To learn cryptographic protocols in a group with individualized keyboards is more motivating. (3, L5, Median 4, Range 1–5, 63%) Mean=3.68, Variance= 1.25.

Additional hypotheses – Preferred scenario

- The individualized concept keyboard is well arranged. (2, B2, 70%)
- The inbuilt chat functionality displays error messages and gives feedback from the other pupils. (3, B2, 88%).

75% voted for the teacher perspective arguing that this role:

- allows for a better exploration of the protocols,
- gives more time to consult the help screens without the time constraints that occur in a collaborative scenario, and
- shows the complete keyboard with all necessary actions in the right order in the start-up configuration.

Further insights

- Find out whether the participants performed better in a posttest organized afterwards than in a pretest before working with the CK. Confirmed!
- 95% agreed that the configuration of the keyboard by drag and drop is quite simple and that they tried to place the keys in a semantically meaningful way
- The contribution of adding sounds or special graphic information to the keys was declined (70% disagreement), but individual shapes of the keys were welcomed by 75% of the participants.

11 Dynamic Interface Reconfiguration within a Distributed Simulation Framework [6]

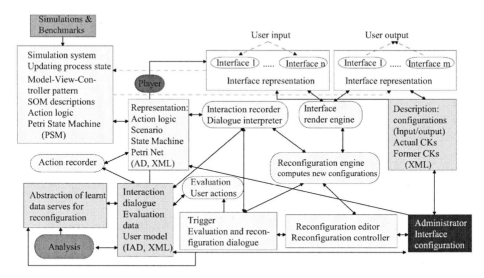

Proposed reconfiguration system architecture

12 Common Work with SRS (Prof. Söffker, [7])

Petrinet-based implementation of a process model – Situation-Operator-Model-based automatic surveillance of human-machine interaction

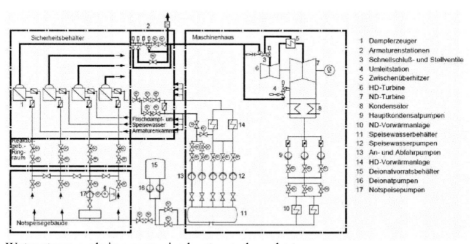

Water-steam-cycle in a pressurized water nuclear plant

Relevant display units – KSG/GfS Simulator Center Essen, Germany

References

1. Collazos, C., Guerrero, L.A., J.A. Pino, J. A.. Ochoa, S. F.: Evaluating Collaborative Learning Processes. Lecture Notes in Computer Science 2440, 2002, pp. 203–221.
2. Cattaneoa,, G., De Santisa, A., and Ferraro Petrillo, U.: Visualization of cryptographic protocols with GRACE. Journal of Visual Languages & Computing Vo. 19 (2), 2008, pp.258-290; doi:10.1016/j.jvlc.2007.05.001
3. Weyers, B.: *Concept Keyboards zur Steuerung und Visualisierung interaktiver kryptographischer Protokolle CoBo'06*. University of Duisburg-Essen (2006)
4. Kováčová, A.: Implementierung des Needham-Schroeder Protokolls in einer verteilten Simulation-sumgebung für kryptografische Standardverfahren. Master's thesis, University of Duisburg-Essen (2007)
5. Selvanadurajan, L.: *Interaktive Visualisierung kryptographischer Protokolle mit Concept Keyboards - Testszenarien und Evaluation.* Master's thesis, University of Duisburg-Essen (2007)
6. Schubert, Chr.: Dynamische Schnittstellenrekonfiguration im Rahmen einer verteilten Simulationsumgebung. Master's thesis, University of Duisburg-Essen (2007)
7. Bousssairi, H.: Petrinetz-basierte Implementierung eines verfahrenstechnischen Prozessmodells – SOM-basierte automatische Überwachung der Mensch-Maschine-Interaktion. Master's thesis, University of Duisburg-Essen (2008)

An error-driven approach for automated user-interface redesign – concepts and architecture

B. Weyers, W. Luther

Department of Computational and Cognitive Science, University of Duisburg-Essen

weyers@inf.uni-due.de, luther@inf.uni-due.de

Abstract. The focus of this talk is the development of a framework for an adaptive user interface based on automatic error recognition and interface redesign. It is well known in Human Computer Interaction research that there is a noticeable gap between the interface and the operation that may be performed by the user and the cognitive representation of the interaction process. Especially the matching process from the user's goal definition to an operation sequence which is executed on an interface is error-prone. Nevertheless, the interface is often to complex and is not able to assist the user in certain way. Our approach presented in this talk combines different research areas to develop a framework which uses automatic generation concepts combined with error-recognition techniques to offer an interface that reduces the gap between user and machine.

Keywords: Interface redesign, adaptive interface, error recognition, automatic interface generation, human machine interaction, cognitive engineering

1 Introduction

Human Computer Interaction (HCI) describes the fundamental problem of interaction as gap between machine's interfaces and the cognition of the human user. Dörner [2] presents different types of errors which can occur by interacting with complex systems. Norman [1, 5] describes the gap in interaction as application of Human Computer Interaction and identifies fundamental problems in the design of interfaces.

Our approach tries to close the gap between the user's cognitive abilities and the real machine state by picking up the understanding of human errors (Dörner) and the way of interface design (Norman) investigated by HCI research [1]. Our idea is to provide a framework for an error driven and automatic redesign of the machine interface during runtime. It should offer continuity between the extremes of no supervision and complete control of any of the user's actions. By automatic redesign the user should be slightly pushed in the correct direction where "correct" should be decided via a suitable distance measure (this is not topic of this talk). The system should assist the user based on the recognition of errors in interaction.

2 Norman's Execution Evaluation Cycle

In [5], Norman introduced a concept of human computer interaction, well known as "Execution Evaluation Cycle" (Figure 1). It describes the way interaction happens between a human user and a machine.

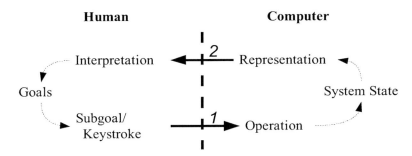

Figure 1: Norman's Execution Evaluation Cycle

Norman identifies two main "gaps" or "gulfs" of interaction. The first one is the "Gulf of Execution" (indicated with *1* in Figure 1). It describes the occurrence of possible errors during the execution of operations. The gulf results from wrong choice of operations by the user, caused by incorrect understanding of operations. The second is the "Gulf of Evaluation" (indicated with *2* in Figure 1) which explains errors that occur during the interpretation of the output. Possible reasons are differing expectations, a wrong interpretation, incorrect or to less information etc. Both gulfs should be minimized to offer a most efficient interaction. We are convinced of the hypothesis that static interfaces cannot reach this requirement of high efficiency[1] in interaction.

3 Reduction of complexity

Every interface (Input and Output) is attended by a reduction of complexity (Figure 2). On the side of Input-Interfaces, complex operations are matched to one or two simple interface widgets, e.g., a single button to start a dish washer. Any time dealing with Output-Interfaces the user is confronted with a reduction of complexity and information of the real system state, e.g., a flashing lamp that indicates a running dish washing process. These reductions may cause (besides other points) Norman's gulfs. The reason for this reduction is the conflicting requirements of (1) offering a complete interface which represents all possible operations or system information and (2) an understandable and interpretable representation which is adapted to the user's limited cognitive abilities.

---

[1] Efficiency here means the ratio between errors and all operations during an interaction.

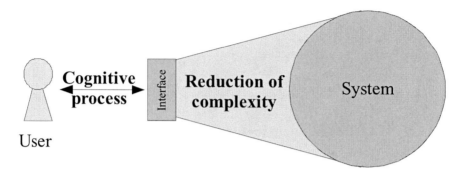

Figure 2: Reduction of complexity in today's interface implementations

4 Proposal

The idea to reduce the gulfs and coincidentally solve the problem of complexity reduction can be summed up in the following points:

- Give the user the possibility to adapt the interface to her wishes
- Create an automatic interface redesign mirroring error recognition and the identification of context changes
- Offer the tools to trigger and manipulate the redesign operations in different roles (as user or administrator).

5 Goals

From this proposal and our idea of an automatic redesign, the following requirements can be established:

- Process- and interaction-orientated redesign of the interface during runtime.
- Reduction of errors in (security sensitive) interaction processes and therefore supervision of the user-system interaction.
- Automatic interface redesign which is transparent and does not distract the user.
- Leave the idea of static interfaces behind and reach a new way of creating high dynamic interfaces which does not generate but reduce errors.

6 Architecture

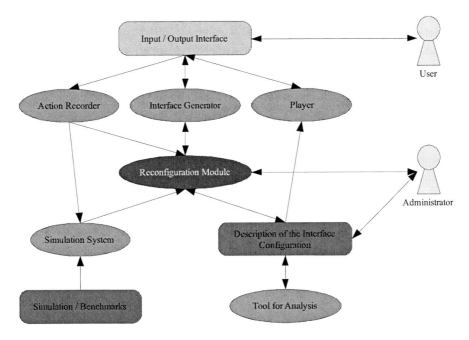

Figure 3: Architecture of an adaptive interface redesign tool inspired by [7]

The architecture (Figure 3) can be separated into (1) Error Recognition and Redesign and (2) System Managing Tools. The latter help to configure the system or connect it to different simulation environments. The Error Recognition and Redesign components stand in the focus of our research.

7 Error Recognition and Automatic Redesign

Figure 4 shows a draft of the main elements involved in the error recognition and redesign component. Mainly three different types of data are used for the error recognition and the redesign process:

1. The **process model** that describes the system (also called action logic).
2. The **user model** that can be separated into two parts:
   a. Interaction record
   b. Occured Errors
3. The set of **design rules** that serve as knowledge for the redesign process.

The Process model can be realized as Situation Operation Model (SOM) [8]. SOM is a meta-modeling approach to structure complex human machine interaction processes. This modeling language assumes that changes in parts of the real world can be understood and also described as a sequence of scenes and actions. In SOM, scenes are modeled as Situations (S) and actions as Operators (O). More information about the SOM approach as a meta-modeling concept can be found in [8].

Based on SOM, error recognition as well as observing tasks can be easily modeled and implemented as shown in [4, 9]. Based on this work, the error recognition component can be realized as extension of the mentioned examples.

The concept of user models is well known from (cooperative) learning and tutoring systems. The user model describes the current user of the system. In the context of a redesign system, errors that occur during interaction as well as in the interaction record, that are parts of the user model. The interaction record describes the former interaction sequence. In SOM notation it would describe the linear sequence of Situations and Operators which occurred in the former interaction process. Out of this information, combined with heuristics and error classification techniques, errors can be recognized. The occurred errors are important for the redesign step in the whole process. Concluding from the occurred error, the redesign component can decide when to start the redesign process and how to redesign the interface. To answer the question how to redesign the interface, design rules have to be generated and involved as second information into the redesign process. Design rules can be generated from well known or newly found patterns and results from usability engineering [10].

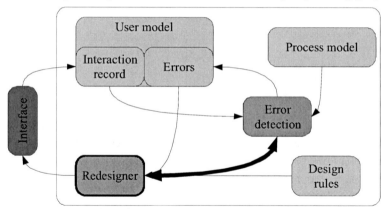

Figure 4: Error Recognition and Redesign Component: An architectural approach

8 Redesign

Figure 5 shows three main questions of automatic interface redesign that we have identified:

1) Who triggers the redesign?
2) When should the redesign be triggered?
3) How should the interface be redesigned?

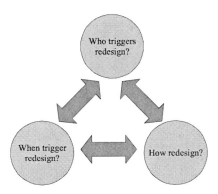

Figure 5: Aspects of Redesign

9 Who should trigger the redesign?

In the context of this question, three different scenarios are possible:

1) The **user** triggers the redesign, if she is dissatisfied with the current interface
2) The **administrator** triggers the redesign, after having evaluated the interaction in a teaching process or having detected deficiencies in the design
3) The **system** triggers the redesign by an error-driven approach.

The third point is in the context of the presented work the central research motivation. The first and second point should also be implemented and play an important role in the context of user acceptance and flexibility (see section 15).

10 When should the interface be redesigned?

The point in time the redesign should start is closely connected to the former described scenarios. In the first two cases, the initialization of the redesign is defined by the user or the administrator. In the third situation, the system has to decide when to trigger the redesign depending on the state of the process. Therefore the question of how to decide automatically when the redesign should start has to be answered. This is part of the current research effort.

11 How should the interface be redesigned?

The underlying decision process of redesign is our main research effort building the basement for a successful redesign. In the context of an error-driven approach the decision process is highly dynamic because of its dependency on an ongoing process of interaction. In [11] a static concept for modeling the action logic of a system was developed in the context of a learning system. As an example for a static concept it is not suitable for our new approach but we gain in experience of modeling the action logic of complex systems.

12 A definition of Extended Petri Nets

In [11] an extension of a basic Petri Net definition is introduced. It was developed in the context of a distributed learning system to teach cryptographic algorithms. Therefore it was necessary to detect students' errors in execution of an algorithm. The new approach was used to model the action logic of the algorithm as well as the behavior of the system in case of errors.

To apply to several implementation issues, two extensions to the basic Petri net definition were introduced. The types of *places* and *transitions* were extended to directly connect them to executable Java Code. Consequently, code is able to be called directly by the execution of transitions and events are triggered by entering tokens in a place. The result is an extended State Transition network (Figure 6).

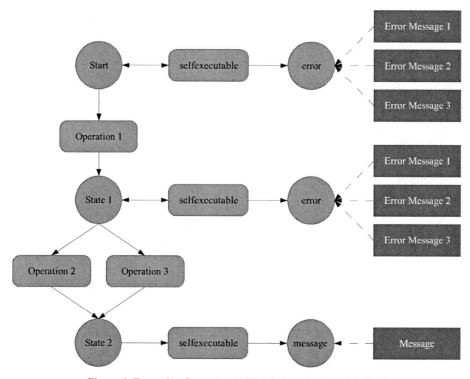

Figure 6: Example of an extended Petri Nets developed in [11]

13   Concept of an error-driven approach

To come back to a dynamic, error-driven approach, Figure 7 gives an idea of a possible decision process. It shows data and information (indicated as ellipses) that are necessary as well as the different components (indicated as rectangles) that are involved in the decision process. Beginning with the collection of different data the interaction will be analyzed concerning the problems that occurred during the interaction. The results will be passed to the decision component to find the best matching design rule for a redesign. The Interface Generator then applies the design rule to the current interface.

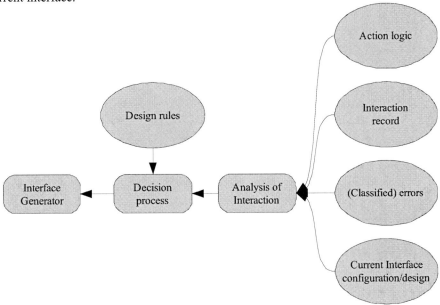

Figure 7: Concept of a dynamic, error-driven approach to interface redesign

14   How to redesign the interface?

How to redesign the interface is the third central question in research of automatic redesign. In this part we are inspired by interaction patterns as well as results from HCI research. Cognitive as well as psychological issues have to be in sight during thinking about design rules for the redesign component. Another important point in this context is how to call the user's attention to a redesign and not cause effects like "Automation Surprises" [6].

15 Problems

From these questions, ideas and concepts described above, many problems result in the scope of this work. The following list should give some ideas which problems are of central interest and have to be solved on condition of the mentioned requirements in section 4 and 5. The list does not intend to be complete.

1. **Ping pong situation**
    The system should offer the redesign initialization to different actors in the scenario (see section 10). In this concept the problem of a "ping pong" situation can occur. In the case of a system's redesign trigger competing against a user's redesign request, a decision has to be made which request the system should follow.
2. **Undermine user's authority**
    The system should never undermine user's authority. It is intended to assist and not to incapacitate the user.
3. **Automation surprises [6]**
    User acceptance is one major requirement and so automation surprises like investigated in the context of aviation techniques is also a major aspect to take care of in the development. A redesign initialized by the system has to be recognizable to the user. The user has to be involved in the redesign process. User centered design [3] is one of the headwords that are important in this context.

16 Perspectives and Outlook

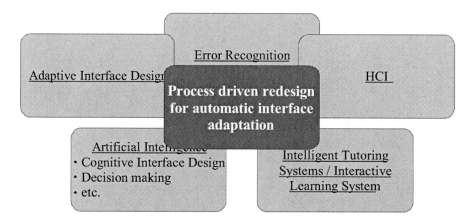

Figure 8: Overview of involved research areas

Figure 8 conveys that this work or system involves many different research fields and results. Our proposal is intended as a highly interdisciplinary research work and involves psychological research, computer science as well as results from Interactive Learning Systems.

17 References

1) A. Dix, J. Finlay, G. D. Abowd and R. Beale, *Human Computer Interaction*. Prentice Hall, 2003.

2) D. Dörner, *Die Logik des Mißlingens*. Rohwolt, Tübingen, 2003.

3) J. M. Flach and C. O. Dominguez, User-Centered Design: Integrating the User, Instrument and Goal. D. Meyer and S Kornblum (Eds.), Attention and Performance XIV. Hillsdale, 1995, NJ:Erlbaum.

4) D. Gamrad, H. Oberheid and D. Söffker, *Formalization and Automated Detection of Human Errors. SICE International Conference on Instrumentation, Control and Information Technology*, August 20-22, 2008, Tokyo, Japan, accepted.

5) D. A. Norman, *The Design of Everyday Things*. B&T, 2002.

6) N. B. Sarter, D. D. Woods and C. E. Billings, *Automation Surprises*. G. Salvendy (Ed.), Handbook of Human Factors and Ergonomics (2nd edition), NY: Wiley, New York, 1997, pp. 1926-1943.

7) C. Schubert, *Dynamische Schnittstellenrekonfiguration im Rahmen einer verteilten Simulationsumgebung*. Diplomarbeit, Universität Duisburg-Essen, 2007.

8) D. Söffker, *Systemtheoretische Modellbildung der wissensgeleiteten Mensch-Maschine-Interaktion*. Habilitationsschrift, Logos, Berlin, 2003.

9) D. Söffker, *Modeling of human errors of the 'man – complex technical system' - interaction: a first qualitative engineering approach*. IMACS World Congress on Scientific Computation, Modelling and Applied Mathematics, Vol. 5, Berlin, 1997, pp. 167-172.

10) J. Tidwell, Designing Interfaces, O'Reilly, Cambridge, 2006.

11) B. Weyers, *Praxisprojekt '06, Architektur und Aktionslogik*, Ausarbeitung. Universität Duisburg-Essen, Duisburg, 2006.

GUI Interaction and Geometric Modeling for Surgical Planning Applications using Superquadrics

R. Cuypers, W. Luther

Institute of Informatics and Cognitive Science, University of Duisburg-Essen

cuypers@inf.uni-due.de, luther@inf.uni-due.de

Abstract. The topics of our talk are advanced human interface interaction and geometrical models that are suited for usage in virtual surgery planning environments. Besides achievements in the branch of speed and accuracy of the anticipated simulation, new interface concepts are presented which allow performing the interaction in a powerful and intuitive way. Concepts for the fully automatic reconstruction of the individual patient's physical structure are discussed along with potential deficits of these approaches that prevent a widespread clinical use. Instead, a semi-automatical system is proposed that allows the user to take control of the reconstruction process when it appears necessary and enhance the quality by providing additional info to the system.

Keywords: GUI, interaction, surgical planning, 3D reconstruction, superquadrics, geometric modeling

1 Introduction

The rise of computer-based operational planning systems resulted in a significant improvement of the quality of orthopedic therapy. Their first aim is to give the surgeon a detailed understanding of the complex interior structure of the patient's body and enable him to select an individualized therapy.

- Decisions to make in operational planning
  - Should a surgery be performed or not?
  - Which material should be used?
  - Which parameter values should be chosen for the surgical procedure to be mechanically (and maybe cosmetically) advantageous?

This talk is mainly concerned with the topic of inserting implants, specifically Total Hip Arthroplasties (THA). The virtual planning of this kind of operation (and many others) requires the manipulation of 3D reconstructed bone shapes. Therefore, models and interfaces are required that allow the surgeon to efficiently and affectively control the involved processes. It partly references our previous work [1], mostly for the superquadric reconstruction part.

2 Understanding Virtual Surgical Planning Systems

- Typical steps of a virtual surgical planning process
  - Reconstruction of the 3D structure of the bone / muscles / tissues
  - Visualization of topology and inherent data (measurements etc.)
  - Simulation of manipulative tasks as resection by different planes etc.

- Conventional Virtual Surgical Planning systems
  - Support reconstruction of relevant bones and tissues from CT / MRI files using flexible but generic geometric models
  - Visualize relevant physiological features
  - Offer limited simulation of manipulation
  - Tasks on the bones
  - Aim for full automatization on all analyzation and reconstruction issues
- Fully automatical reconstruction often unfeasible due to lack of human intuition of computers
- However, operational palling requires exactness in the sub-millimeter category
  - Therefore a semi-automatical approach is favored
    - The user gets full feedback for all the algorithm steps
    - The user always has the option to provide additional
    - information that improves the processing results
- Generic geometrical models provide few information about the bone's features and are usually very complex
  - Generic surface models
    - Triangle models
    - B-Spline surfaces
    - Disadvantages of these models
      - Large number of parameters
      - Quality of fit can only be computed at high expense
      - Complicated to handle within a graphical user interface
    - Therefore, we use implicit superquadric models.

3 Bone Decomposition and Recovery

The reconstructional part consists in manually decomposing the bone's point cloud into subparts using a special selection tool and fitting superquadric-based bone models to the individual parts.

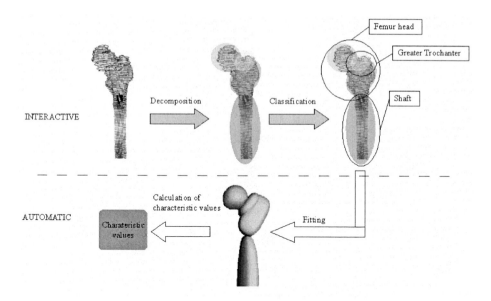

Fig. 1. Process-flow for the bone reconstruction part.

4 Superquadric Shapes

Superquadrics are a family of geometrical shapes that are defined by the implicit equation [2]:

$$\left(\left(\frac{x}{a_1}\right)^{\frac{2}{\varepsilon_2}} + \left(\frac{y}{a_2}\right)^{\frac{2}{\varepsilon_2}}\right)^{\frac{\varepsilon_2}{\varepsilon_1}} + \left(\frac{z}{a_3}\right)^{\frac{2}{\varepsilon_1}} = 1 \tag{1}$$

- Parameters
  - (x, y, z) : Position of a point that shall be tested against the SQ
  - (a_1, a_2, a_3): Scaling of the superquadric
  - $(\varepsilon_1, \varepsilon_2)$: Roundness of the superquadric
  - Additional parameters like origin, orientation etc.

- Advantages of superquadrics
  - High descriptive power with few parameters
  - Well-behaved inside-outside function
  - Easily extensible using deformations
  - Can produce more complex models by coupling multiple SQs
  - Simple enough for fast user interaction

- These properties make superquadrics useful for surface-modeling, approximation and gui-based manipulation

To support the reconstruction process, the calculation of several superquadric features has been implemented.

- Computation of radial distance between points and superquadrics
- Computation of absolute distance between points and superquadrics

- Reconstruction of superquadric geometry
  - Fitting superquadric to point cloud
    - Basic model
    - Deformable model [3]

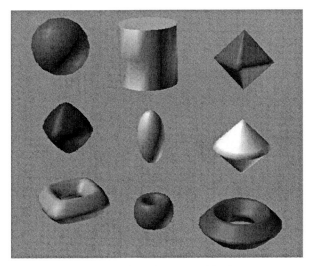

Fig. 2. A selection of superquadric-shapes.

5 Superquadric-Based Bone-Modeling

- Model the individual components of each bone, the implants and the crafting tools relevant for the surgery using superquadrics
- Use inherent superquadric capabilities to process geometrical operations very fast

- Full lower limbs modeled
  - Femoral bone
  - Tibial bone
  - Pelvis
  - Foot

- THR implants modeled [4]
  - Aesculap BiCONTACT_S NK510T
  - Aesculap NK115T

- Cutting tool modeled

- The cutting steps on the bone can be simulated using CSG (Constructive Solid Geometry) operations

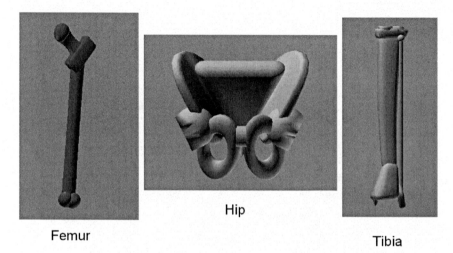

Hip

Femur

Tibia

Fig. 3. Superquadric-based bone models.

6 Recovery of Bone-Geometry using Superquadrics

Once the relevant point-data for every bone part has been determined, a recovery-process is started that generates a fitted superquadric for every bone component. The fitting superquadric for a bone is the one whose surface has minimum radial distance to its point cloud. This comes down to a nonlinear optimization problem that involves minimizing the expression [3], [5], [6]:

$$
\min_A \sqrt{a_1 a_2 a_3} \sum_{i=1}^n \left(f(x_i, y_i, z_i)^{\varepsilon_1} - 1 \right)^2 \tag{2}
$$

where f is the radial distance between a point and the superquadric surface.

- Fully automatic bone geometry
  - Problem: Fully automatic recovery produces often dissatisfying results
  - Reason: Most shape decomposition algorithms are generic and dissect the bone disadvantageously

- Observation: Intuitive manual decomposition of humans may be very different
- Conclusion: The user needs a way to interact with the recovery process!

- Other reasons for bad recovery results
  - Bad input data (gaps, outliers, wrong area of interest)
  - Unknown target-shape conditions (see bone classification section)

- What we want
  - The decomposition of the points should be in a way that the parts can be easily approximated by the superquadric model

- What we need
  - A selection tool that can adapt itself to all possibly needed superquadric shapes

- So why not use the superquadric model for selection?

→ Create superquadric-based selection tool

7 Superquadric Selection Tool

In terms of the semi-automatical reconstruction approach, the user has to manually determine the relevant point data for each individual bone part. Since the components have been shown to be efficiently representable by superquadrics and the reconstruction process comes down to a series of superquadric-fits, it makes sense to use a superquadric-shaped selection-tool for the bone-decomposition part. Conceptually, the proposed tool is an extension of the *SoDragger*-Interface of the Coin3D-API [7]. The user translates, rotates and scales a number of geometrical shapes whose parameters are directly projected onto those of the underlying superquadric. This system is used in twofold ways: First, as a template for point grouping, where only the points lying inside the transformed superquadric are affected by the selection procedure. Second, as a manipulator for every already existing superquadric-shape in the scene. The latter allows for example correcting the superquadric-model of a reconstructed bone-part if necessary.

Fig. 4. The superquadric selection tool.

- Specific Goals
  - Allow the selection of points by superquadric-shape using the mouse
  - Allow the manipulation of reconstructed superquadric-shapes
- Supported models
  - Superquadrics
  - Extended deformable superquadrics (Solina) [3]

8 Classification Profile for Bones

In order to determine if a reconstructed bone can be considered as valid, a classification profile has been developed using inherent superquadric properties.

- Bone attributes
  - Length, curvature, etc.
  - Decomposition into subshapes
  - Connectivity, size, position etc. of the subshapes compared to each other

- Properties of the bone parts
  - Every bone part can be represented by a certain superquadric type
  - Every superquadric is defined by a limited set of parameters
  - A valid bone part can therefore be defined by intervals of parameters

- Bone profiles
  - Every bone part has a profile consisting of minimum, maximum and initial-guess for fitting parameters
  - If the bone type and possible abnormalities are known beforehand, the user can assign a profile to the relevant point set that helps the recovery process

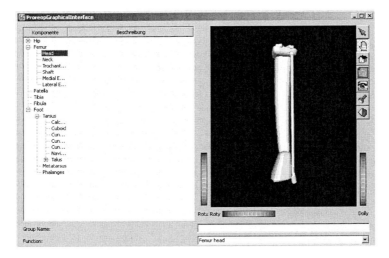

Fig. 5. GUI-based bone-classification.

9 Bone Feature Extraction

After the bones have been fitted, certain bone-features can be directly extracted from the underlying superquadric-model. Due to the superquadrics' capabilities, the measurements can be conducted very efficiently.

femurAxis	Origin	0.241	0.0625	0.121
	Direction	0.87	0.226	0.438
femurHeadCenter		0.022	0.085	0.137
femurHeadDiameter		0.0527		
trochanterHeadPosition		21.57	49.77	159.66
minMarrowDiameter		0.024		
kneeAxis	Origin	0.433	0.076	0.112
	Direction	0.0083	0.991	-0.129
kneeZylinderA		0.0684		
kneeZylinderB		0.0613		
femurNeckDiameter		0.0396		

Fig. 6. A selection of bone-features.

10 Implementation

An implementation of the before mentioned concepts has been done that roughly covers the steps from importing the bone point-data to the generation of the composed superquadric-model of the bone. The user is presented a GUI that includes four viewports to present the data, several widgets for loading and saving data and triggering the individual steps of the reconstruction process. Inside the viewports, the superquadric manipulation tool is shown. There are buttons in the toolbar that allow for the import of bone point-sets stored as textfiles on the harddisk. After a point-set has been loaded, the surgeon can utilize the superquadric selection tool to select points and compile them into a group. Grouped components are assigned a different color from the rest of the data. After assigning all the point-groups, one can select any of them and start a fitting process on it that spawns a superquadric matching the points in the group. Before starting the individual optimization process, it is possible to name the corresponding bone part so the initial and boundary parameters are appropriately set. When done with his work, the user can store the whole scene and reload it at a later time.

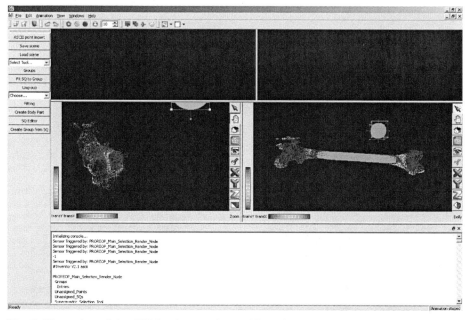

Fig. 7. Illustration of the GUI-implementation. White points are unassigned, dark-gray points have already been grouped. The bright-gray shapes are fitted superquadrics and the spherical shape with the wire-box is the sq-based selection tool.

11 Further Applications

There is a multitude of possible applications for the above mentioned in computer-guided medical applications.

- Possible applications
  - Surgical planning environments
  - Fracture modeling
  - Bone feature comparison
  - Automatic generation of training sets for surgical simulators

12 Further Work

- Generate a model of the whole human lower-limb apparatus
- Add more prosthesis and cutting tool models
- Add implementations of more sophisticated superquadric models
- Increase the level of automatism
- Connect all the features to a fully-guided operational-planning process.

References

1. R. Cuypers, Z. Tang, W. Luther und J. Pauli, Efficient and Accurate Femur Reconstruction using Model-Based Segmentation and Superquadric Shapes , *Proceedings of the IASTED International Conference on Telehealth and Assistive Technologies* 2008
2. A.H. Barr, Superquadrics and angle-preserving transformations. *IEEE computer graphics and applications,* 1:1, 1981, pp. 11–23.
3. A. Jaklič, A. Leonardis and F. Solina, *Segmentation and recovery of Superquadrics.* Vol. 20 of *Computational Imaging and Vision.* Kluwer, Dordrecht, 2000.
4. Aesculap AG Surgical instrumentation manufacturer *http://www.aesculap.com/*
5. L. Chevalier, F. Jaillet, and A. Baskurt, Segmentation and superquadric modeling of 3D objects. In *Journal of Winter School of Computer Graphics, WSCG'03,* 11:2, Feb. 2003, pp. 232–239.
6. F. Banégas, M. Jaeger, D. Michelucci, and M. Roelens, The ellipsoidal skeleton in medical applications. In *Proceedings of the sixth ACM symposium on solid modeling and applications,* ACM Press, 2001, pp. 30–38.
7. Coin3D Scenegraph Library   *http://www.coin3d.org/*

Modeling the Human-Machine-Interaction in the Context of Knowledge-Guided Behavior

Dirk Söffker

Chair of Dynamics and Control, University of Duisburg-Essen, D-47048 Duisburg

1 What is modeling?

Modeling I-IV

Modeling fixes the constants and fix relations in

- the real world,
- dynamic changes, and
- other relevant and context-related structures,

independent of the modeling technique. Modeling is not necessarily connected to the explicit usage of modeling techniques. Assuming this statement to be true, graphic relations denoting structural contexts and/or relations are also models as human recognition and memory of previous and/or former experiences and the like. As Wittgenstein [1] stated, people do not check every day to validate the existence of stairs; it is useful to work with such mental constants -in this case, to make assumptions based on what we know about the constant behavior of an individual's environment, which will be referred to below as the individual's outside world.

Humans need mental models to reduce the complexity of their interactions with the outside world, assuming constants in the structure of the outside world, which allows them to predict the future based on those known and experienced constants of the known structure. Modeling human-machine interaction (HMI) assumes that human interaction behavior here is understood as human cognitive-driven, or knowledge-guided, behavior. It should be noted that other human interactions with parts of or the entire outside world can be described successfully with frequency domain I/O relations (such as the stimulus response behavior between physical signals for the eyes and the muscle response realizing speed up / brake or lane change maneuvers) as well as sequentially realized or repeated behavior (like making coffee) with algorithms [2]. The human interaction to be considered here is knowledge-guided human interaction based on the use of the human cognition by cognitive functions, procedures and routines. Cognition assumes awareness and the human ability to perceive, store, handle, and combine information and to use information for actual and/or upcoming human actions, including human planning and reflections. Rasmussen distinguishes three levels of Human-Machine-Interaction (HMI): it is clear the control develops methods and approaches to describe the underlying constants using and developing suitable approaches. The open question remains: which approach can be used to describe the knowledge-based human behavior?

2   What is interaction? What are typical examples?

Interaction / Human-Machine-Interaction I-VI Interaction

includes the interaction of two units, call them systems. But what is a system? Assuming that the systems to be considered are acting autonomously, humans can also be understand as systems. Besides technical system-system interaction and social science-oriented human-human interaction, possible combinations also include human-system interaction, which is considered here. Descriptions of human-human interaction are usually based on description techniques using linguistic terms, while technical system-system interaction can be described by formalized, typically qualitative and/or quantitative mathematics-based approaches. But how can human-machine interaction be described? As an example, let us consider the interaction between human operators guiding the trains at the main station of Hagen railway station in Germany. The human operator sits in the control center in front of a set of monitors. On the monitors, the actual state of the rails and crossings is graphically displayed. In front of the operator, a special keyboard forwards operator input to the operating system, enabling the operator to switch crossings, signs and the like, preparing the rail routes for the trains. These considerations are based on a study conducted by the author together with several students, in which the human-machine interaction at this electronic operating center of the Deutsche Bahn AG was examined from 1995 to 1997. Details and a list of further reading are given in [3].

The interaction itself is examined with respect to clustering, analysis of human errors and formalization. It was observed that the formalization is strongly related to the internal structure of the sequences, whereby this structure is related to the logic of the action sequence. This observation forms the core of the situation-operator model extended and developed by the author. In contrast to previous publications about situations and/or states and actions, this examination [3] leads to the internal structuring of situations and requirements related to the modeling of actions, including the one defined as situation-operator sequence or modeling (SOM). Other examples of the HMI class under consideration are the pilot-aircraft and the operator-power plant relationship.

The typical feature of the HMI-systems to be considered here are

- the complexity of the technical system to be controlled,
- the spatial distinction between the guiding operator and the system,
- the high grade of automatic or autonomous subsystems to be guided by the operator, and
- the guiding, observing and/or monitoring character of the human operators work.

The closed loop of control scientists as a metaphor shows the principal structure of the interaction, whereby, in contrast to technical control loops, the human-system interaction closed loop cannot be described by mathematical equations or by technical/physical values. Furthermore, the question arises of how the higher goals of technical control loops, such as stability, robustness, observability and controllability, can be examined within this context. To solve the question about the suitable description of human-machine interactions, it is necessary to determine what information is observable from the scenes that can be used for measuring within the modeling context. But what is a measurement? Within this context, spoken words and visible actions on the part of the human as well as displayed parameters and observable behavior on the part of the machine that can be measured or monitored are denoted as measurements. Furthermore, knowledge is available about the context of actions.

3 Modeling and models of human cognition

Models of human cognition, Comparison of the modeling approaches

Cacciabue [4] introduced criteria to compare models of human cognition. Human cognition in general includes cognitive functions and cognitive procedures, whereby cognitive functions include the phenomenological observable aspects of human learning, human reasoning and planning. Cognitive procedures relate the functions with the memory and so forth. The criteria developed by Cacciabue [4] are PIPE, AoR, and KB, denoting respectively perception, interpretation, planning and execution, allocation of resources and knowledge base. Unfortunately it is not clear in which way these aspects of human models are able to detail all possible aspects of human cognition modeling as well as the implementation aspects of human models for programming. This will be discussed below and is detailed in the publication [5]. The models are discussed in detail by Cacciabue and will not be dealt with in detail again here.

4 New and advanced requirements for human cognition modeling

Requirements for HMI-modeling I-II

From the modeling perspective, HMI models have to represent the necessary relations between human cognition and the human environment, to clearly describe the internal organization of cognition, and to represent the known learning, reasoning, planning and decision-making abilities of humans. It is necessary for them not only to explain known behavior, but also to predict human behavior in specified situations in a constructive manner. Furthermore, successful models should include previous models and their abilities. The need for a minimum modeling depth that allows not only the description of behavior but also the construction of future behavior is even more advanced. From this point of view, models cannot be pure textual models only; they have to be implemented in a reproduction scheme that makes the models independent of human understanding and interpretation. These requirements lead to the idea of a programmable model of human cognition, able to realize observed human-cognitive-oriented interaction. Such a model can be called a validated model if the implemented (programmed) model is able to work for special situations in the same way as humans. From the HMI point of view, it seems to be necessary for the relation between the outside world structure and the internal organization and storage of knowledge is known and clear. Also, the models must be able to be applied to various real situations.

In conclusion, HMI modeling should

- describe the structure of things happening in the real world (changing of facts) connecting in the minimum two units (the human and the environment; two systems, like plant and controller) in a very general way,
- connect modeling techniques for the outside world (classical modeling approaches) with the model of the memory of human cognition,
- be able to represent human learning behavior and therefore to analyze the structure of human errors, and
- provide much more detailed models than the models introduced above.

5 The Situation-Operator-Model modeling technique

Qualitative Modeling I-IV

The core of this approach is the assumption that changes in the parts of the real world under consideration are understood as a sequence of effects described by the item's scenes and actions (of the real world RW). In the proposed approach, the definition of the item's scenes and actions are coordinated in a double win. They are related to each other and relate the assumed structure of the real world to the structure of the database-called the mental model-of the intelligent system. Intelligent systems and humans (human operators) are included in the real world. Depending on their principal sensory inputs, their natural or technical perceptions, and the related knowledge base, intelligent systems and the like adapt and learn only parts or aspects of the real word. These parts are modeled using the developed situation and operator calculus. The describable part is called a system.

The item situation, which is time-fixed and system- and problem-equivalent, is used to describe the internal system structure (as a part of the RW). Here, only the logical structure of the 3D-space, time and function-oriented connections are of interest. The item operator is used to model effects / actions changing scenes (modeled as situations) in time. The situation S consists of characteristics C and a set of relations R. The characteristics are linguistic terms describing the nodes of facts (as perceptible qualities). This will include physical, informational, functional and logical connections. To describe the relations r_i known problem-related modeling techniques, like ODEs, DAEs, algorithms or more general ones, even graphical illustrations (such as Petri-nets) can be used. The SOM-approach only gives the framework for modeling the structure of changeable scenes and therefore maps the reality of the real world to a formalizable representation using the proposed structural framework. This is useful for describing problems where the system structure is complex (and cannot be described with available [single] approaches) and cannot be modeled using single approaches. This is also useful for describing interactions between human operators and their environment.

The introduced item characteristic C also includes the possibility of representing time-dependent parameters P as an example. The complete set of relations R (of Cs) fixes the structure of the real world scene under consideration modeled as situation S. The situation concept introduced consequently allows the integration of different types of engineering-like descriptions.

The illustrated item operator is used for modeling a) internal (passive) connections of situations and b) changes between situations.

The operator O is understood and modeled from a functional point of view: the operator is an information-theoretic term which is defined by its function F (as the output) and the related necessary assumptions. Here explicit and implicit assumptions eA, iA are distinguished. The function F will only be realized, if the explicit assumptions eA are fulfilled. The iA includes the constraints between eA and F of the operator. The eA are of the same quality as the characteristics C of S. For the internal structure of the operator other descriptions textual, logical, mathematical or problem-related descriptions. The double use of the term operator O is graphicly illustrated.

The description of complex systems using a Situation-Operator model allows

- the mixture of different types of (variable) quantities (the relations R can be different ones within the situation S),
- the integration of logical and numerical quantities (by different characteristics C), and
- the description of real-world problems using a mixture of a complex set of descriptions (variables).

Operators are used to model the system changes (changes of situations). This defines the events of the change of the considered part of the real world, the system. Operators and situations are closely connected due to the identity (partial or complete) of the characteristics of the situations and the explicit assumptions of the operators. This includes the situation consisting of passive operators (internal causal relation: because), whereby the change is carried out by active operators (external causal relation: to), shown in Figure 2. The change in the considered world results as a sequence of actions modeled by operators as illustrated in Figure 4. It should be noted that operators correspond to situations. Both are used not only for structural representation of the system's organization but also for internal representation and storage of human operators and intelligent systems. They are the core/background of all the higher organized internal (cognitive) operations and functions of the IS, like learning and planning and also of the proposed supervision concept. Furthermore, it becomes clear that learning results from a loop between estimated behavior (using cognitive structures with elements of the mental model for such purposes as planning) and the observed behavior of the environment, which-in the case of non-agreement with the prediction-provides good reasons to reflect on the assumed behavior and its elements, which is called learning. This loop also makes clear i) that learning without feedback (or without environment) is not possible and ii) that different points, elements and facts are necessary for successful learning processes, including previous knowledge, available cognitive functions (like dealing with mental elements and planning procedures), the possibility of suitable environment reaction observation and also the ability to observe and remark differences between postulated and observed behavior. These differences can be easily detailed and used to structure the learning process as well as possible difficulties within this complex loop. The author discusses in detail several student theses regarding a) classification of human errors, b) modeling of human learning, c) reconstruction of mental models from interviews, d) reconstruction of mental models from observed and monitored human behavior during disasters and e) classification of dynamic group decisions in [3].

6 Resulting SOM cognitive modeling approach

HMI modeled SOM-approach, HMS-SOM I-II, Resulting structure, Facts

The resulting approach appears as a framework assuming a structure of facts and behaviors on the part of the outside world side; these can be modeled using the SOM approach. A copy of the known effects and behaviors are within the human brain as the basis for cognitive functions and procedures. The human-machine interface appears (in addition to the ergonomic interface) to be the logical interface between the human's inside view and the outside world of reality and thus to be the situation trajectory in time. The proposed approach appears to be a system-theoretic approach realizing aspects of observability and controllability. Details are given in [3]. The use of models allows the prediction and reconstruction of nonmeasurable states. This assumes that the model and the described system are identical. The resulting structure of the proposed cognitive modeling approach is as follows: Signals from reality are perceived by the perception filter. The filter is strongly defined by the previous knowledge of the individual mental model. The key features perceived by the perception filter are called characteristics and are identified by identical elements within the mental model. The interaction between mental model and perception filters leads to so-called scene awareness on the part of the human. Using mental functions, the elements of the mental copy of reality are used for mental operations such as preparing actions. Depending on mental and physical abilities, the planned (and previously simulated) actions are realized and used to change the human's environment (the outside world) in the desired way. In contrast to classical control loops, here qualitative/quantitative descriptions are used to describe the I/O-relation. As the controller of the situation, The cognitive acting human carries out the interactions in a situative, problem-equivalent and goal-oriented manner to attain desired interactions and goals. The feedback is typically designed by the designer of the working situation; in abnormal situations, interacting humans design the feedback interactions themselves. This includes learning, concluding and acting and reflects the unique flexible reactions of humans.

7 Concept of a situative, flexible supervision module

Example: Explainin the SOM-idea in detail -Structuring Complexity of Interaction

The concept of a supervision module is also based on the Situation-Operator-Model (SOM). The structuring of the underlying logic of action can be represented formally and also can be formalized using situation vector and related database-oriented operators. The formal check detects in a first step situation-based the logic of the action in relation to the actual situation.

The example of car changing lanes or passing maneuvers demonstrates the concept as well as the SOM-concept in general. In this case, no relations are specified. Data compression is carried out by extracting the parameters of characteristics from the time-discrete and partly continuous raw sensor data. In the module SSetting up situation", different algorithms are used to calculate the parameters of the characteristics, which have discrete values and are event discrete.

For example, to calculate the parameters of a characteristic lane change, a driver uses the distances and velocities of the vehicles in front and behind in the lanes on the left and right. The sample algorithms provided illustrate only the extraction process and may be replaced by more sophisticated algorithms, such as neural networks and fuzzy logic, to model the structure of the scene at a higher abstraction level. The possible actions of the driver are modeled as operators, which are stored in the basis operator library. From the set of basis operators, such as brake, drive and accelerate, meta-operators are constructed, like normal lane change to right and fast passing. For each operator, the assumptions used to apply it are denoted on the left side, while its effects are stated on the right side, distinguishing whether this change is certain (solid line) or probabilistic (dashed line). The proposed approach in combination with the example demonstrates the automated supervision of human drivers. The main advantage of this concept is that the uniform formulation leads to a modular and easily extendable system due to the structuring by the different abstraction levels. Full integration of the proposed automated supervision would lead to an autonomous system that can perform the supervised task on its own. Furthermore, the process of learning can be modeled using this technique, which is not limited to describing the example presented here but can be used to describe the interaction of a wide rage of autonomous systems. Hence, a learning system based on the SOM description could also respond to a changing environment; such a system is currently in development.

8 Autonomous systems as the next step in realizing SOM-based interactions

Next Goal, Example, Autonomous Robots, Implementation, Realization

In the field of dynamics and control the SOM approach is used to develop and implement the goal-driven autonomous behavior of technical systems in a first step toward autonomous robots. Therefore, the concept introduced above is programmed to implement cognitive functions. In this way the implementation of the cognitive model appears as cognitive control. The goal of the proposed autonomous robot is to react situation-based and additionally flexible so as to interact in a goal-oriented fashion with unknown environments; the implementation of this goal is based on a client-server concept. The proposed cognitive feedback approach encompasses the abilities of both learning and adaptation, and is -like human behavior- able to produce situative, problem-dependent and goal-oriented behavior. The assumption is that the outside world of the controller is formalizable (that the SOM approach can be applied) and in this way constant (independent from any kind of mathematical characterization, like linear or nonlinear). This kind of approach produces a cognitively controlled technical system.

9 Literature

1. Wittgenstein L: Philosophical Investigations (PI). In: Anscombe GEM, Rhees R (eds.), Anscombe GEM (trans.), Oxford, Blackwell, 1953.
2. Söffker D: From human-machine-interaction modeling to new concepts constructing autonomous systems: a phenomenological engineering-oriented approach. Journal of Intelligent and Robotic Systems 32, 191–205, 2001.
3. Söffker D: Systemtheoretische Modellbildung der wissensgeleiteten Mensch-Maschine-Interaktion. Logos Wissenschaftsverlag, Berlin, 2003.
4. Cacciabue PC: Modelling and Simulation of Human Behavior in System Control. Springer, London, 1998.
5. Söffker D: Cognitive Approaches realizing flexible Interaction behavior of Cognitive Technical Systems: a Comparison. Proc. 5th IMACS Symposium on Mathematical Modeling, Vienna University of Technology, Austria, 8 pages, February 6-8, 2006.
6. Edwards E: Introductory overview. In: Wiener EL, Nagel DC (Eds.): Human factors in aviation. San Diego, Academic Press, 1988.
7. Rouse WB: Models for Human Problem solving: detection, diagnosis, and compensation for system failures: Automatica 19 (6), 613-625.
8. Rasmussen J, Pedersen AM, Schmidt K: Taxonomy for cognitive work analysis. In Rasmussen J, Brehmer B, de Montmollin MDe, Leplat J (Eds.): Proc. 1st MOHAWD Workshop, Liege, May 15-16, Vol 1 ESPRIT Basic Research Project 3105, EC, Brussels, Belgium, 1–153, 1990.
9. Reason J: Human Error. Cambridge University Press, Cambridge, UK, 1990.

10. Sheridan TB: Telerobotics: Automation and Human Supervisory control. The MIT Press, Cambridge, USA, 1992.

11. Hollnagel E: Human reliability analysis: context and control. Academic Press, London, 1993.

12. Dörner D: Die Logik des Misslingens, Verlag Rowohlt, Reinbek 1989.

13. McCarthy J, Hayes PJ: Some philosophical problems from the standpoint of artificial intelligence. In: Meltzer B, Mitchie D, Swann M (Eds.) Machine Intelligence (4), Edinburgh University Press, Edinburgh, 463–502, 1969.

14. Söffker D: Closing loops: a Unified View from Control to Information Science. Proc. 4th IMACS Symposium on Mathematical Modeling, Vienna University of Technology, Austria, February 5-7, 8 pages, 2003.

15. Söffker, D.: Interaction of Intelligent and Autonomous Systems - Part I: Qualitative Structuring of Interactions. MCMDS-Mathematical and Computer Modelling of Dynamical Systems, Vol. 14. No. 4, 2008, pp. 303-318.

16. Ahle, E.; Söffker, D.: Interaction of Intelligent and Autonomous Systems - Part II: Realisation of Cognitive Technical Systems. MCMDS-Mathematical and Computer Modelling of Dynamical Systems, Vol. 14. No. 4, 2008, pp. 319-339.

Modeling of the
Human-Machine-Interaction

in the context of the human
knowledge-guided behavior

Dirk Söffker
Email: soeffker@uni-due.de

Chair of
Dynamics and Control'

University of Duisburg-Essen
Campus Duisburg

Outline

- **Modeling**
  Fundamentals, basic ideas

- **Human-Machine Interaction**
  Interaction > Cognitive models > SOM-modeling > Examples

- **Unified View to Dynamical Systems**
  Technical control, human-machine system, autonomous system
  From signals to systems
    ⇔ From physical values to data
    ⇔ From technical control to autonomous interaction

* Systemtheoretic View * Examples * Vision

Söffker, D.:
Modeling Human-Machine Interaction
© for all figures/illustrations by SRS U DuE

Modeling I

What is "modeling" (from the engineering point of view)?

- sratching for 'knowledge' about ...
- 'make a model' describing relations ... and use it later
- combing former known information/models or assumed axioms ...
- ...

Advantages of modeling

- fix the information and store it ...
- use the model ... but for what?
  > predict behavior(s) based on previously known ... (models)
  > to combine it ... and produce new information/models ...
- ...

Typical modeling approaches

- differential equation / transfer function / state space models
- neural net / fuzzyfication ¸ behavior description
- algorithm ...
- ...

Verification / validation

Söffker, D.:
Modeling Human-Machine Interaction
© for all figures/illustrations by SRS U DuE

Modeling Ib

What is "modeling" (from the engineering point of view)?

...

Advantages of modeling

...

Typical modeling approaches
- differential equation / transfer function / state space models
- neural net / fuzzyfication / behavior description
- algorithm ...

Verification / validation

- Verification: follows the rules, 'works as specified'
                 in identity with former knowledge/information
- Validation: proofed on the base of experiments

Söffker, D.:
Modeling Human-Machine Interaction
© for all figures/illustrations by SRS U DuE

Modeling II

Invariants of modeling approaches

- modeling the constants > What are invariants, what are the constants?
- constants in dynamical behaviors are (underlying) structures/relations
- constants are related to modeling/description techniques
  <>problem to be described vs. modeling approach

Modeling of complexity/ of complex behaviors

- is difficult (to do, to verify, to validate esp. due to the 'complexity'
- is typically based on the modeling of elements/modules etc.
- starts with assumptions to reduce partially the complexity
- can elementary be validated /
  can be macro-oriented phenomenologically be validated

Purpose of modeling

- understand what may be inside / the effects / the relations
- used for further appl. (control / supervision > model-based appr.)
- ...

Modeling IIb

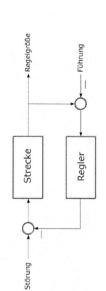

Modeling III (Examples from the control/system view)

Causality ⇔ From the cause to the effect

| Cause | | Effect |

i) time-based sequence
ii) internal connections

Base of the consideration I: System

Base of the consideration II: Effect

technical systems:
> physical values
>> physical effects

Modeling IIIc

Complex connections / complex systems

> Connected complex thinking
⇔ Modeling of complex connections

> Prediction of complex time behavior

>> ‚Limits of the Growth' (Club of Rome, 70ties)

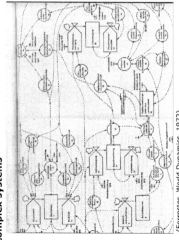

(Forrester, World Dynamics, 1972)

Modeling IV

What is "human-machine interaction"?

Different levels of human behavior / human action / human interaction

> Rasmussen, 197x/8x:
  - skill-based level   >> frequency-domain base
  - rule-based level   >> algorithms
  - knowledge-base level   >> ?

> Focus on knowledge-guided / cognitive behavior

Human- Machine Interaction I

Example: Supervision and control of railway traffic

here:
Hagen Electronic Operating Center of the Deutsche Bahn AG

(Söffker, 2001)

Interaction — Two-way reaction between players or systems

At minimum two "systems" are interacting:

System – System
Human – Human
Human – System

> **Human-machine interaction**
> **Human-machine system**

What is the interaction?
How can the interaction be described?

Human- Machine Interaction II

Example: Supervision and control of railway traffic

here:
Hagen Electronic Operating Center of the Deutsche Bahn AG

Other examples

(Söffker, 2001)

Human-Machine Interaction III

How can the visible, measurable and audible interaction be modeled?

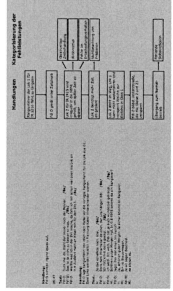

(Clauß [Thesis]; Söffker, 1996)

Söffker, D.:
Modeling Human-Machine Interaction
© for all figures/illustrations by SRS U DuE

Human-Machine Interaction IVa

Abstraction and aggregation lead to a strong reduction of interaction elements

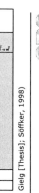

(Gielg [Thesis]; Söffker, 1998)

Söffker, D.:
Modeling Human-Machine Interaction
© for all figures/illustrations by SRS U DuE

Human-Machine Interaction IVb

Grafische Darstellung der Handlungssequenz:

Hypothese / Plan:

$S_{A1} \rightarrow O_{V1} \rightarrow S_1 \rightarrow O_{V1} \rightarrow S_{R1}$

neu: $S_{A1} \rightarrow O_{V1} \rightarrow S_1 \rightarrow O_{V1} \rightarrow S_{R1}$

Tatsächlicher Verlauf:

(Gielg [Thesis]; Söffker, 1998)

Söffker, D.:
Modeling Human-Machine Interaction
© for all figures/illustrations by SRS U DuE

Human-Machine Interaction VI (and assumptions)

Causality ⇔ from cause to effect

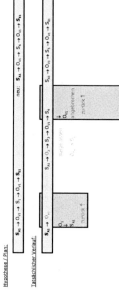

Which is the adequate description? (techn./physic. values > information)

Higher goals: - stability / dynamics - robustness - observability
- controllability - > automatic control

Söffker, D.:
Modeling Human-Machine Interaction
© for all figures/illustrations by SRS U DuE

i) Final chain
ii) Inner connections

Modeling of Human Cognition I

What is cognition?

- Related to human

    thinking,
    learning,
    making conclusions,
    learning,
    planning,
    reasoning,
    ....

Söffker, D.:
Modeling Human-Machine Interaction
© for all figures/illustrations by SRS U DuE

Comparison of Modeling Approaches

Table 18   Overall matching of the six models of cognition with the requirement of the Reference Model of Cognition.

Requirement	Edwards SHEL (1973)	Icone IIPS (1983)	Rasmussen SI/SRK (1986)	Reason Fallible Machine (1990)	Sheridan Supervisory Control (1992)	Hollnagel COCOM (1993a)
1. Perc./Interp./Plan./Exec. (PIPE)	*	***	***	*	***	***
1.1 Connections within PIPE	* Seq.	*** Seq.	*** Seq.	*	*** Cyc.	*** Cyc.
1.2 Option of Cont./Back on PIPE	**	–	–	–	*	***
2. Allocation of Resources (AoR)	*	***	***	***	*	***
2.1 Competition of AoR and PIPE	–	***	***	*	*	***
2.2 Connections of AoR and Memory/KB	**	–	*	***	***	**
2.3 Effects of Cont./Back on AoR	–	*	*	**	*	***
3. Memory/KB (KB)	–	*	*	***	***	**
3.1 Connections between KB and PIPE	–	*	**	*	*	**

The following measures apply:
- Not treated
-- Implicit consideration
Modeled at minimal level   *** Modeled in detail   Seq. Sequential   Cyc. Cyclic

(Cacciabue, 1998)

Söffker, D.:
Modeling Human-Machine Interaction
© for all figures/illustrations by SRS U DuE

Which is the adequate description? (techn./physic. values > information)

Higher goals: - stability / dynamics   - robustness   - observability
- controllability   - > automatic control

What can be done to model HMI using measurable values (I/O)?

- Save A/V-data
- Save systems state / system states
- Save inputs (technical data)
- Save outputs (technical data)

> Relate the I/O-relation!

(but: relation between the two sides has to be considered)
(>>Lets talk about the human)

Söffker, D.:
Modeling Human-Machine Interaction
© for all figures/illustrations by SRS U DuE

Modeling of Human Cognition I

Basic functionalities of human cognition

- cognitive functions
    > can be realized in realtime (in the so-called working memory)

- cognitive procedures
    > is connected with the knowledge base/memory

⇒ Criteria for modeling human cognition (Cacciabue, 1998)

PIPE (Perception, Interpretation, Planning, and Execution)

AoR (Allocation of Resources)

KB (Knowledge Base)

Söffker, D.:
Modeling Human-Machine Interaction
© for all figures/illustrations by SRS U DuE

Requirements for HMI Modeling I

From the modeling point of view
- Model should be able to be used to explain, to predict, ...
- Micro- or Macro- ...
- Model should be in the minimum be verified ...

From the HMI point of view
- Model should be connected to
    i) the known behavior of the system
    ii) the measurable behavior of the human
- Model should integrate available, related knowledge and experiences
- Should avoid any kind of psychologically-oriented interpretations!

Conclusions
- Modeling the constants/invariants of the interaction appears as the core.
- Modeling must include more detailed approaches (> metamodelling appr.)

UNIVERSITÄT DUISBURG ESSEN

Söffker, D.:
Modeling Human-Machine Interaction
© for all figures/illustrations by SRS U DuE

Requirements for HMI Modeling II

Acceptable assumptions
- Event discrete consideration
- Not to be able to predict Human behavior (# ethic aspect)

Purpose of the HMI modeling
- Understand HMI based on a formal description
- Be able to have a look inside human interaction without interpretations
- Be able to copy human flexibility (situative, awareness, ...)

Solution ideas
- Use situation calculus (McCarthy) and detail it
- Use action schema and detail it
- Relate situation and action on this base together etc.
- Map outside world to internal model
- Use understanding of cognition and realize a cognitive machine

UNIVERSITÄT DUISBURG ESSEN

Söffker, D.:
Modeling Human-Machine Interaction
© for all figures/illustrations by SRS U DuE

Qualitative Modeling Approach I
(structural variable systems)

Situation:

The term situation describes a fixed problem constellation and denotes the considered system.

The situation consists of an inner structure, which also allows the integration of time-variant values.

The graphical representation is realized by characteristic (C) and inner relations (R). Different detailed graphical representations are possible.

(Söffker, 1998f, 2001, 2003)

UNIVERSITÄT DUISBURG ESSEN

Söffker, D.:
Modeling Human-Machine Interaction
© for all figures/illustrations by SRS U DuE

Qualitative Modeling II
(structural variable systems)

Operator:

Operators are used to represent functional connections of real world facts. The connection can be passive (constitutional) or active ('ability to change something'). Operators represent/model outer world facts.

The function of an operator is denoted by (F). The explicit and implicit assumption for realization of F (eA, iA) are used as 'input'.

For detailed modeling known techniques will be used. The SOM technique is working as a metamodeling approach.

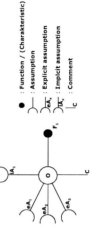

- ● : Function / (Charakteristic)
- : Assumption
- eA : Explicit assumption
- iA : Implicit assumption
- c : Comment

(Söffker, 1998f, 2001, 2003)

UNIVERSITÄT DUISBURG ESSEN

Söffker, D.:
Modeling Human-Machine Interaction
© for all figures/illustrations by SRS U DuE

Qualitative Modeling III
(structural variable systems ⇔ cognitive systems ⇔ e.g., HMS)

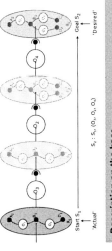

Start S_1
'Actual'

$S_1 : S_1, (O_1, O_2, O_3)$

O_3 O_2 O_4

Goal S_2
'Desired'

**Description on the base
of assumed facts about the real world**

Necessary:
- **Model of the 'world' > Human > 'Human interaction'**
- **Controllability**
- **Observability**

Söffker, D.:
Modeling Human-Machine Interaction
© for all figures/illustrations by SRS U DuE

UNIVERSITÄT
D U I S B U R G
E S S E N

Sequence of Actions / (Operators) Changes Situations:

Start S_1
'Now'

$S_1 : S_1, (O_1, O_2, O_3)$

O_3 O_2 O_4

Goal S_2
'In the future'

- looks like an algorithm
- allows modeling of changing situations (> HMI / interacting IS)
  > also allows modeling of structural changes

Söffker, D.:
Modeling Human-Machine Interaction
© for all figures/illustrations by SRS U DuE

UNIVERSITÄT
D U I S B U R G
E S S E N

HMS – SOM I

**System theoretical aspects of
Human-Machine Interaction:**

- Description   >> SOM approach
- Observability   >> strongly depends on the
              mental model and the
              cognitive abilities
- Controllability   >> strongly depends on the
              mental model and the
              cognitive abilities
- Automatic control   ??
- Stability   ??

Söffker, D.:
Modeling Human-Machine Interaction
© for all figures/illustrations by SRS U DuE

UNIVERSITÄT
D U I S B U R G
E S S E N

HMS
–
SOM II

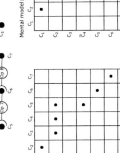

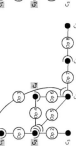

Mental model:

Reality:

(Söffker, 2001)

Söffker, D.:
Modeling Human-Machine Interaction
© for all figures/illustrations by SRS U DuE

UNIVERSITÄT
D U I S B U R G
E S S E N

Resulting Structure from the Cognitive Modeling Point of View

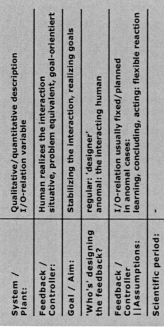

Söffker, D.:
Modeling Human-Machine Interaction
© for all figures/illustrations by SRS U DuE

Human-Machine Systems: Facts

System / Plant:	Qualitative/quantitative description I/O-relation variable
Feedback / Controller:	Human realizes the interaction situative, problem equivalent, goal-orientiert
Goal / Aim:	Stabilizing the interaction, realizing goals
'Who's' designing the feedback?	regular: 'designer' anomal: the interacting human
Feedback / Controller / Assumptions:	I/O-relation usually fixed/planned in anomal cases: learning, concluding, acting; flexible reaction
Scientific period:	-

Söffker, D.:
Modeling Human-Machine Interaction
© for all figures/illustrations by SRS U DuE

Example: Conception of Automated Supervision

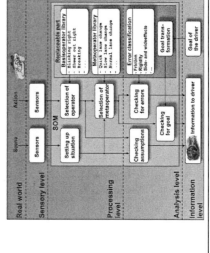

Söffker, D.:
Modeling Human-Machine Interaction
© for all figures/illustrations by SRS U DuE

Example: Explaing the SOM-idea in detail - Structuring Complexity of Interaction

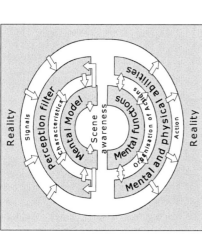

Söffker, D.:
Modeling Human-Machine Interaction
© for all figures/illustrations by SRS U DuE

Supervision Automat
Sensory level

- Detection of scene and action by sensors

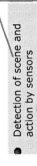

UNIVERSITÄT DUISBURG ESSEN

Söffker, D.:
Modeling Human-Machine Interaction
© for all figures/illustrations by SRS U DuE

Supervision Automat
Processing level

- Setting up situation as characteristic vector
- Selection of operator
  - Operator library
  - Selection of metaoperator
  - Metaoperator library

UNIVERSITÄT DUISBURG ESSEN

Söffker, D.:
Modeling Human-Machine Interaction
© for all figures/illustrations by SRS U DuE

Representation of Traffic Situation
Characteristic vector

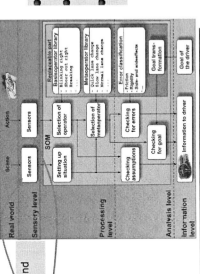

Characteristic	Parameter
actual lane	[1, 2, 3, ...]
actual velocity	[real]
turn signal set	[right/left/no]
passing lane exists	[no/right/left/ right and left]
actual lane free	[yes/no]
lane change possible	[no/right/left/ right and left]
acceleration possible	[yes/no]

- Characteristic vector represents situation
- Fixed structure of characteristics
- Parameters change during passing maneuver
- Specification of parameters according to traffic situation

UNIVERSITÄT DUISBURG ESSEN

Söffker, D.:
Modeling Human-Machine Interaction
© for all figures/illustrations by SRS U DuE

Extraction of Characteristics
Example

- Determination of information of the scene by sensors
- Processing of sensor data by algorithm, Fuzzy logic, ...
- Complex scene represented as situation by characteristic vector

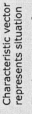

Sensor data

Velocity	v_w	[real]
Velocity	v_x	[real]
Velocity	v_y	[real]
Velocity	v_z	[real]
Distance	S_{xy}	[real]
Distance	S_{yz}	[real]
Distance	S_{xw}	[real]
Distance	S_{zw}	[real]
Velocity	v_v	[real]

Data compression

$$S_{ey} = S_{ey}(S_y + S_{yw} + S_{yz})$$
$$S_{ex} = S_{ex}(S_x + S_{xw} + S_{xz})$$
$$S_{ez} = S_{ez}(S_z + S_{zw} + S_{zz})$$

if $((S_{ey} > 0) \wedge (S_{ez} > 0))$ then
return right and left
else if $((S_{ey} > 0) \wedge (S_{ez} > 0))$ then
return left
else if $((S_{ey} > 0) \wedge (S_{ez} > 0))$ then
return right
else return no

Situation

Characteristic

- actual lane
- actual velocity
- turn signal set
- passing lane exists
- actual lane free
- lane change possible
- acceleration possible

Parameter

- [1, 2, 3, ...]
- [real]
- [right/left/no]
- [no/right/left/ right and left]
- [yes/no]
- [no/right/left/ right and left]
- [yes/no]

UNIVERSITÄT DUISBURG ESSEN

Söffker, D.:
Modeling Human-Machine Interaction
© for all figures/illustrations by SRS U DuE

Modeling driver's actions

Basisoperators

Basisoperator library

Sheer to right

Brake

Accelerate

...

- Drive
- Accelerate
- Decelerate
- Brake
- ...

Söffker, D.:
Modeling Human-Machine Interaction
© for all figures/illustrations by SRS U DuE

UNIVERSITÄT
D U I S B U R G
E S S E N

Basisoperator - Example

Example

Sheer to right

Assumptions
- One parameter or
- many parameters of a characteristic possible

Function
- Changes parameter of a characteristic
- Changes many parameters

Söffker, D.:
Modeling Human-Machine Interaction
© for all figures/illustrations by SRS U DuE

Modeling driver's actions

Metaoperators

Metaoperator library

Normal lane change to right
Set turn signal right - Sheer to right - Set turn signal back - Drive

Fast lane change to right

...

- Fast lane change to left/right
- Slow lane change to left/right
- Fast passing
- Slow passing
- ...

Söffker, D.:
Modeling Human-Machine Interaction
© for all figures/illustrations by SRS U DuE

UNIVERSITÄT
D U I S B U R G
E S S E N

Metaoperator

Example

Normal lane change to right
Set turn signal right - Sheer to right - Set turn signal back - Drive

Söffker, D.:
Modeling Human-Machine Interaction
© for all figures/illustrations by SRS U DuE

Assumptions and function result from Basisoperators

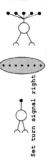

Set turn signal right

Sheer to right

Set turn signal back

Drive

Söffker, D.:
Modeling Human-Machine Interaction
© for all figures/illustrations by SRS U DuE

Supervision Automat

Analysis level

- Checking assumptions
- Checking for goal
  - Goal transformation
- Checking for errors
  - Errors in SOM

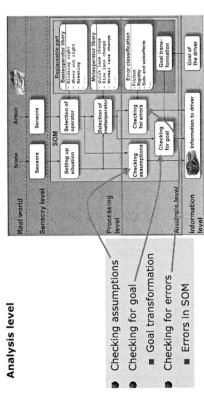

UNIVERSITÄT
DUISBURG
ESSEN

Söffker, D.:
Modeling Human-Machine Interaction
© for all figures/illustrations by SRS U DuE

From Scene to Situation

Example

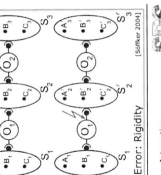

Scene

actual lane	[2]
actual velocity	[100 km/h]
turn signal set	[no]
passing lane exists	[left and right]
actual lane free	[yes]
lane change possible	[no]
acceleration possible	[no]

Situation

Accelerate

Actual operator

UNIVERSITÄT
DUISBURG
ESSEN

Söffker, D.:
Modeling Human-Machine Interaction
© for all figures/illustrations by SRS U DuE

Automated Supervision

Operators
with respect to
assumptions
- Drive
- Decelerate
- Brake

Operators
with respect to goal
1. Drive
2. Decelerate
3. Brake

Scene

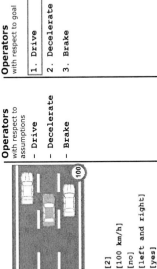

Situation

actual lane	[2]
actual velocity	[100 km/h]
turn signal set	[no]
passing lane exists	[left and right]
actual lane free	[yes]
lane change possible	[no]
acceleration possible	[no]

Goal ⟶ „Drive safely"

UNIVERSITÄT
DUISBURG
ESSEN

Söffker, D.:
Modeling Human-Machine Interaction
© for all figures/illustrations by SRS U DuE

Modeling Human Error in SOM

Classification according to Dörner

- Establishment of fix goals
- Rigidity
- Side- and wideeffects
- Magic hypothesis
- Central reduction
- No part-goal elaboration

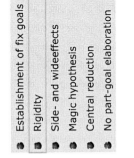

Error: Rigidity

[Söffker 2004]

UNIVERSITÄT
DUISBURG
ESSEN

Söffker, D.:
Modeling Human-Machine Interaction
© for all figures/illustrations by SRS U DuE

Next Goal: Autonomous Systems

(Situative reaction must be flexible for interaction with unknown environments.)

(Ahle, Söffker, 2005)

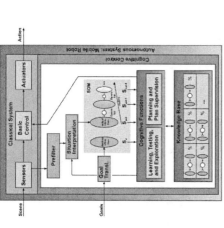

Implementation

Client/Server Concept:

- Client requires prefiltered situations
- Choice of operators within cognitive learning and planning routines
- Client knows about the actual action performed
- Experiences will be saved in a database

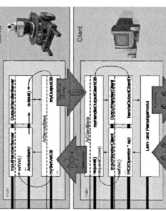

(SRS: Ahle, Olle, 2005)

Autonomous Robots

(Goal: situative, flexible, goal-oriented interaction with unknown environments)

Sensors
Video camera
Laser Scanner
Sonar Sensors

Actuators
Wheels/Motors
Gripper

(Söffker, 2001;
Ahle, Söffker, 2005)

Realization
Test setup

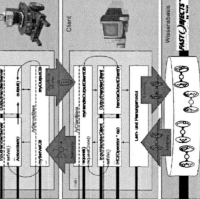

- Two given goals
- Given interpretation and operators
- 22 experiences
- 4 autonomous generated meta-operators

Cognitive Technical System – Autonomous System

System / Plant:	Unknown environment, I/O-structure variable
Feedback / Controller:	AS is able to learn and adapts itself: situative, problem equivalent, goal-oriented
Goal / Aim:	Realization of given goals
'Who's designing the feedback?	Abstract: the designer In detail: the AS (learn, plan, observe)
Properties of Plant / Controller || Assumptions:	'World' is formalizable, cognitive routines plus SOM formalization allows learning, sensors/actuators equivalent
Scientific period:	21st century (> cognitive technical systems)

Söffker, D.:
Modeling Human-Machine Interaction
© for all figures/illustrations by SRS U DuE

UNIVERSITÄT DUISBURG ESSEN

Summary and Outlook

- **From Control to Human-Machine Systems and Autonomous Systems**
  System-theoretic view to systems dynamics possible
  SOM approach as a metamodeling approach

- **Dynamical Systems**
  Key: Use of knowledge about the structure of interaction
  HMI can be described based on a cognitive modeling approach
  Human flexibility can be modeled and 'realized' by autonomous systems

- **What's new?**
  - Modeling approaches (ODE > Information science–oriented approaches)
    >> Control goals and aims are becoming more abstract
    >> Complexity may be modeled
    Core: system-theoretic foundation

- **Vision:** > formalizable learning capabilities will affect control approaches <

Söffker, D.:
Modeling Human-Machine Interaction
© for all figures/illustrations by SRS U DuE

UNIVERSITÄT DUISBURG ESSEN

Realization

Test in lab environment

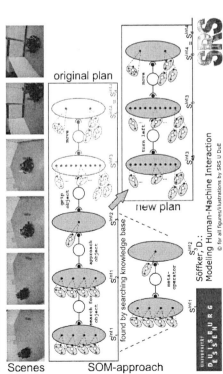

original plan

new plan

found by searching knowledge base

Scenes   SOM-approach

Söffker, D.:
Modeling Human-Machine Interaction
© for all figures/illustrations by SRS U DuE

UNIVERSITÄT DUISBURG ESSEN

Literature

Ahle, E.: Autonomous Systems: A Cognitive-Oriented Approach Applied to Mobile Robotics, Dissertation, Shaker 2007.

Cacciabue, P.C.: Modelling and Simulation of Human Behaviour in System Control. Springer 1998.

Dörner, D: Die Logik des Misslingens, Verlag Rowohlt, Reinbek, 1989.

Helander, M., et al: Handbook of Human-Computer Interaction. Elsevier, 2nd edition, 1997.

Russel, Norvig: Artificial Intelligence – A Modern Approach. Prentice Hall, 1995.

Söffker, D.: Systemtheoretische Modellbildung der wissensgeleiteten Mensch-Maschine-Interaktion. Post-Doctoral Thesis, Bergische Universität – Gesamthochschule Wuppertal 2001, also with Logos Verlag, Berlin, 2003.

For the papers by Ahle and Söffker:
www.srs.uni-duc.de > Personen > Söffker > Literatur
resp.: 2 Journal Papers in Computer and Mathematical Modeling of Dynamical Systems, 2008.

Söffker, D.:
Modeling Human-Machine Interaction
© for all figures/illustrations by SRS U DuE

UNIVERSITÄT DUISBURG ESSEN

Simulation of Human-Machine-Interaction for the Detection of Human Errors

Dennis Gamrad, Hendrik Oberheid, Dirk Söffker

University of Duisburg-Essen, Chair of Dynamics and Control
German Aerospace Center, Institut of Flight Guidance
{dennis.gamrad, soeffker}@uni-due.de, hendrik.oberheid@dlr.de

1 Simulation of SOM using software tools for high-level Petri Nets

Similarities between SOM and HPNs, Simulation of SOM-based models, Tools for high-level Petri Nets, Overview about different patterns

The construction and analysis of SOM-based models has formerly been realized in the form of symbolic, graphical representations [1] or implementations in higher (textual) programming languages [2]. To combine the advantages of a graphical representation with the benefits of formal executable models [3] studies the possibility and benefits of developing specialized net patterns based on existing formalisms and tools for high-level Petri Nets (HPNs) to simulate and analyze selected properties of SOM-based models. Two different formalisms and tools for HPNs are considered for SOM namely the software CPN Tools, for the simulation of Coloured Petri Nets [4] and the software Renew for the simulation of Reference Nets [5].

The motivation to consider HPN formalisms for the simulation of SOM-based models is due to a number of reasons. First of all, certain structural similarities exist between SOM-based models and HPN formalisms. They are both inherently bipartite formalisms consisting of an active element representing the system functionality and a passive element representing the system state. In this respect a natural and intuitive correspondence exists between the notion of operators in SOM and transitions in HPN formalisms. The meaning of the terms situation in SOM and of places (in combination with a token) in HPNs is also closely related, depending partially on the concrete HPN formalism.

It should be noted that in a number of features the flexibility and openness of the SOM approach as a conceptual framework and meta-modeling technique goes significantly beyond what is covered by existing HPN formalisms and allows model behaviors, which can not be realized and simulated by existing HPN tools. Some of the limitations (especially with regard to hybrid, discrete-continuous processes and fixed net topology) and some possible extensions to deal with them are discussed in [6].

In the following the realization of Net-Patterns, which represent SOM-based models, is described. The description starts with the presentation of net patterns for discrete SOM-based models, which are extended to hybrid (discrete-continuous) SOM-based models.

1.1 Net-Patterns for discrete SOM-based models

Patterns for discrete SOM-based models

Two different basic net patterns are briefly presented in this section, which define a correspondence between the elements situation, operator, characteristic, relation, and assumption of an SOM-based model and the terms place, transition, arc and token of a HPN model. In a first step, relations R and operators O are assumed to be time-invariant, so similar patterns can be realized with most discrete HPN formalisms, including Coloured Petri Nets and Reference Nets.

In patterns I, the situation S is represented by a place, containing a single token. The token is a data structure potentially consisting of different data types, representing the characteristics C of the situation S. Each individual operator O is modeled by a transition which can change the values of the characteristics. The assumptions iA and eA for the execution of a certain operator O_i are validated by the guard function and the actual bindings of the transition.

In Pattern II, the situation S is also represented by a place, containing a single token with the characteristics of the Situation S. In contrast to pattern I, a transition represents a group of operators with similar structure and function while the exact quantitative effects of the individual operator are specified in an operator token.

1.2 Net-Patterns for hybrid SOM-based models

Patterns for hybrid SOM-based models

Continuous dynamics and time-variant relations between characteristics can be added to the pattern for discrete SOM-based models by 1) connection of the discrete HPN model with an external continuous model e.g. via TCP/IP or 2) integration of 'active' tokens with autonomous dynamics directly into the HPN model. The common idea of both methods is that a continuous process is running in principal independently and for a limited time autonomously of the HPN software.

The second method can be realized in Renew due to the special reference net formalism of Renew. In [3], the patterns for discrete SOM-based models are extended in such a way to allow whole java threads as tokens representing the SOM situation S. Thus, a continuous computation of situation characteristics C according to the time-variant relations R of the situation becomes possible. The threads run autonomously besides Renew and are synchronized by common time objects while Renew can continuously query the situation characteristics to test if the assumptions are fulfilled to apply an operator O to the current situation S.

2 Classification and formalization of human errors

Automatic detection of human errors, Human errors in complex dynamical systems

According to Dörner, human errors are classified with respect to the interactions of humans in complex dynamical systems [7,8]. A classical action approach is assumed, which appears as a control loop between the environment and the human operator. Coming from psychology word models are used to describe and distinguish different human errors, which are divided into the four main clusters goal elaboration, decision processing, control of the actions, and errors due to internal cognitive organization problems. The different error types were demonstrated on various examples from different application fields.

To formalize human errors according to Dörner's classification the SOM approach can be used. As an example the error 'rigidity' is considered and visualized with SOM notation on slide 8 [1,9]. In the situation trajectory depicted in the lower part of the figure, the desired situation S_2 is not reached as planned in the upper part, due to external effects and disturbances. Instead, a different and unexpected situation S_{2a} is resulting. Given the new situation, O_2 will not lead to the desired goal due to the changed structural situation $S_{2a.}$ The human error 'rigidity' includes that the known and previously planned operator O_2 is nevertheless realized inconsiderately by the human operator, although the assumptions for its application no longer hold.

3   State-Space-based Detection of Human Errors

State-space-based detection of human errors I – II

State spaces of Coloured Petri Net models are discrete, directed graphs (digraphs). The digraph contains as nodes of the graph different states (markings), which the model may obtain, starting from a certain initial state and given the models constraints. The arcs of the graph represent possible transitions between these states. Adopting an appropriate implementation technique and net patterns to model SOM in CPN Tools as presented in [6], the states are interpreted as SOM-situations while the state transitions are associated with SOM-operators.

State spaces can be generated automatically in CPN tools. Depending on the properties of the model and available computational resources, the resulting state spaces may be complete (representing all reachable states) or only partial (representing a subset of all reachable states e.g. within a certain search depth with regard to the initial condition). Accordingly, it becomes possible to automatically determine a (partial or complete) set of reachable SOM-situations and calculate possible SOM-operator-sequences between these situations.

Using the calculated space of reachable situations is a valuable approach for the investigation of human errors in Human-Machine-System (HMS) interaction, since it allows relating the observed actions of the human operator to the (known) set of reachable situations and executable operators of the system. On the basis of the state space, it becomes possible to formally determine desirable (goal) and non-desirable (unsafe) states/situations reachable by the HMS and then check if the human operator's actions serve to approach (come closer to) a certain goal situation. It is also possible to observe if an action sequence serves to pursue one single goal consistently or iterates/jumps between various competing goals etc.

The automatic detection of human errors within the interaction of the system is realized through formal state space query functions programmed in CPN Tools. In order to keep the functions reusable for different kinds of Human-Machine Systems, the queries are built in a manner that the structure of the error detection is generic and remains the same independent of the specific system. Only the concrete definitions of what desirable/non-desirable/ goal situations mean in a certain application context are system specific and have to be exchanged.

4 Realization of Error Detection

4.1 Structure of the simulation environment

Structure of the simulation environment

The programmed implementation of the proposed concept is realized by an experimental environment, consisting on the one hand of a simulated or real technical process controlled by a human operator and on the other hand of a software tool for high-level Petri Nets (slide 11). The human operator interacts with the technical process over an appropriate graphical user interface. Via a TCP/IP communication all information about state changes in the simulated process (whether user-induced or autonomous) are passed on from the technical process to the CPN Tools simulator. Communication over this connection is unidirectional in the sense that the CPN simulator receives information from the process, but no information is transferred back from the CPN model to the process.

The CPN model contains a copy/reproduction of the relevant parts of the technical process, which is structured according to the SOM approach. By constantly receiving information about each process state change, the CPN simulator tracks the entire situation trajectory of the process. Using the state space functionalities of CPN Tools the space of reachable situations is computed, starting from a certain initial situation. The observed situation trajectory and the observed operations of the user are then mapped into that state space and related the set of the situations, which would have been reachable from a given situation in the trajectory.

4.2 Modeling of Human-Machine-Interactions within an example process

Example: Arcade game application, Modeling of interactions between human and game

In this contribution, the considered process is an arcade style game (like 'Boulder Dash' or 'Sokoban') [10] providing a graphical interface to a human operator and a TCP/IP communication with CPN Tools. Within this formalizable synthetic environment an agent has to be controlled within a grid-based environment. The agent is able to perform six different kinds of actions. The environment consists of static and dynamic elements with different behaviors combinable to complex scenarios using the integrated level editor. In general, the task of the human operator consists of first picking up a certain number of 'emeralds' and then finishing the level by leaving the scenario through an exit door .

In the current development state the arcade game was chosen, due to its simple handling by editing own levels and custom elements. Nevertheless, the structure of the experimental environment and the developed functions for error detection are not specific to the arcade game.

The possible interactions within the arcade game are modeled using patterns for SOM-based models. Slide 13 illustrates a part of the interaction model, which represents the moving actions of the agent. The moving actions of the monsters are also related to the same place, but defined on a separate model page. The functions of the transitions are detailed on subpages on a lower hierarchical level. The situation is modeled by a place consisting of one token, which is a set of data types. In the whole model, each of the agent's actions as well as each of the independent acting monsters' actions is represented by one transition, which is linked to the place. By firing of a transition the corresponding operator is performed and the situation and token on the place respectively is changed.

4.3 Error Detection Model

Error detection scheme, Detection of the human error rigidity

This section describes two CPN model pages designed to detect errors in the control sequence of the operator. The error detection is based on an analysis of the state space/situation space.

The first page named 'ErrorDetectionModel' (slide 14) provides a structure for the interference of 'possible user goals' on the basis of the observed action sequence of the human operator. An objective definition of 'goal situations' in this application encloses the set of situations, which represent a successful completion of the level. However, it is not known beforehand, which of the potential 'goal situations' (ways of completing the level) the operator will actually be pursuing during each stage of the simulation. A subset of all objective 'goal situations' is thus identified and denoted as 'possible user goal(s)'. These 'possible user goals' are inferred from the observed actions of the operator by checking if these actions serve to come nearer to that goal. The 'possible user goals' are the basis for the following formal error detection realized on the subpage 'RigidityDetection'.

Here, a transition and submodule for the detection of the specific human error 'rigidity' is connected. The subpage 'RigidityDetection' with the functionality behind the transition is shown on slide 15. The page contains two transitions ('RigidityPossible', 'RigidtyImpossible') to decide if rigidity may have been the cause for an erroneous human action in the last move. On the one hand, the transition 'Rigidity Impossible' will always fire if the list of possible user goals is not empty. That is, the last user action effectively served to pursue/come nearer to a valid goal situation in the interaction and thus cannot be termed an error of 'rigidity'. On the other hand, 'rigidity possible' will fire if the list of possible user goals remained empty. In this case, the function 'list_poss_rigid' will further check if an observed action, which is now inadequate would have been adequate to realize any of previously 'possible user goals' in the situations observed before (using tokens stored on the place 'previously possible user goals'). If this is the case, it is concluded that 'rigidity' may have occurred and the previous goals, which may have caused the misguided actions, are recorded on the place 'rigidity candidate goals'.

4.4 Results

Example: Simulation

The automated detection of the human error 'rigidity' was implemented and tested for the arcade game application introduced in section 4.2. Some results are shown by an example interaction within the presented scenario (slide 12). The whole sequence of interactions is visualized on slide 16 representing a subset of the entire state space generated on the SOM-based model. The ellipses represent situations, which are linked by the operators. The occurred situations are red-colored and the direct successors are grey-colored.

References

1. Söffker, D.: Systemtheoretische Modellbildung der wissensgeleiteten Mensch-Maschine-Interaktion. Logos Wissenschaftsverlag, Berlin (2003) – auch: Habilitationsschrift, Bergische Universität Wuppertal, Deutschland (2001)
2. Ahle, E., Söffker, D.: Interaction of Intelligent and Autonomous Systems - Part II: Realization of Cognitive Technical Systems. MCMDS-Mathematical and Computer Modelling of Dynamical Systems, Vol. 14, No. 4, 319-339 (2008)
3. Gamrad, D.: Entwicklung von Mustern höherer Petri Netze zur rechnergestützten Simulation und Analyse von Situations-Operator-Modellen. Diploma thesis, University of Duisburg-Essen, Chair of Dynamics and Control, (2006)
4. Jensen, K.: Coloured Petri Nets. Basic Concepts, Analysis Methods and Practical Use. Vol. 1-3, Springer-Verlag (1997)
5. Kummer, O.: Referenznetze. Logos Wissenschaftsverlag, Berlin (2002)
6. Gamrad, D., Oberheid, H., Söffker, D.: Supervision of Open Systems using a Situation-Operator-Modeling Approach and Higher Petri Net Formalisms. Proc. of 2007 IEEE Int. Conf. on Systems, Man and Cybernetics, Montréal, Canada, 925-930 (2007)
7. Dörner, D.: Die Logik des Mißlingens, Strategisches Denken in komplexen Situationen. Rowohlt Verlag (2002)
8. Schaub, H.: Modellierung der Handlungsorganisation. Verlag Hans Huber, Bern, (1993)
9. Söffker, D.: Understanding MMI from a system-theoretic view - Part I and Part II. Proc. 9th IFAC, IFIP, IFORS, IEA Symposium Analysis, Design, and Evaluation of Human-Machine Systems, Atlanta, Georgia, USA (2004)
10. Rocks'n'Diamonds Website, http://www.artsoft.org/rocksndiamonds/

Simulation of Human-Machine-Interaction for the Detection of Human Errors

Dennis Gamrad, Hendrik Oberheid*, Dirk Söffker

Contact: dennis.gamrad@uni-due.de
Website: www.srs.uni-due.de

*DLR, Institute of Flight Guidance

Chair of Dynamics and Control
University of Duisburg-Essen

Simulation of SOM-based models

- Special symbolic and higher programming languages
- Simulation and analysis of SOM-based models
- Tools for SOM-based models not available
- Structural similarities between SOM and high-level Petri Nets (HPNs)
- Tools for HPNs available: CPN Tools and Renew

Which HPN-patterns are suitable for the simulation and analysis of SOM-based models?

Similarities between SOM and HPNs

SOM-based models and high-level Petri Nets consist of static und dynamic elements.

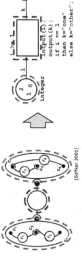

[Söffker 2001]

Tools for high-level Petri Nets

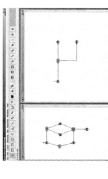

CPN Tools

- Coloured Petri Nets [Jensen 1997]
- State space analysis
- Extendable

Renew

- Referencenets [Kummer 2002]
- No automatic analysis
- Highly extendable

Overview about different patterns

HPN-patterns for SOM-based models

Hybrid (discrete – continuous) SOM-based models

Discrete SOM-based models

Static operators   Dynamic operators   External programs   Java-threads

The patterns can be combined and hierarchized.

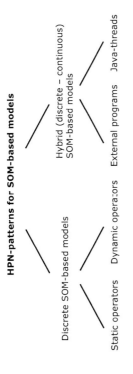

Gamrad, Oberheid, Söffker: Simulation of Human-Machine-Interaction for the Detection of Human Errors
© for all figures/illustrations by SRS U DuE

4

Patterns for discrete SOM-based models

Static operator

$(c1,c2,...)$

Dynamic operator

$(i,m,h,p),$
$...$

$(c1,c2,...)$

Gamrad, Oberheid, Söffker: Simulation of Human-Machine-Interaction for the Detection of Human Errors
© for all figures/illustrations by SRS U DuE

5

Patterns for hybrid SOM-based models

Time object

Thread 1
Thread 2
• • •
Thread n

Renew

Net classes

Synchronization   Continuous process   Discrete process

Gamrad, Oberheid, Söffker: Simulation of Human-Machine-Interaction for the Detection of Human Errors
© for all figures/illustrations by SRS U DuE

6

Automatic detection of human errors

Role of human operators

- Supervisory control [Sheridan, 1992]
- Flexibility and creativity used as fallback level
- Human error rate
- Time pressure, stress level, and system complexity

http://www.wikipedia.de

Gamrad, Oberheid, Söffker: Simulation of Human-Machine-Interaction for the Detection of Human Errors
© for all figures/illustrations by SRS U DuE

7

Human errors in complex dynamical systems

Classification according to Dörner

- Interactions in complex dynamical systems
- Described by word models
- Formalization using SOM approach

Formalization of the error rigidity

- External effects and disturbances
- Previously planned operators no longer goal directed
- Strategy not changed and planned goal not reached

Other human errors are formalized in [Söffker, 2001].

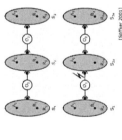

State-space-based detection of human errors I

CPN Tools for Coloured Petri Nets

- Automated state space generation
- Usage of formal query functions

State space of Coloured Petri Nets

- Discrete, directed graph (digraph)
- Nodes represent states (markings)
- Arcs represent transitions

State space of SOM-based models

- Nodes interpreted as SOM-situations
- Arcs associated with SOM-operators

State-space-based detection of human errors II

Formal query functions on a generated state space

- Application independent by a generic structure
- Only detailed definitions are system specific

Fundamental queries to analyze human behavior

- Operators and interaction model (state space) as input
- Reachable situations and executable operators
- Desirable (goal) and non-desirable situations

Structure of the simulation environment

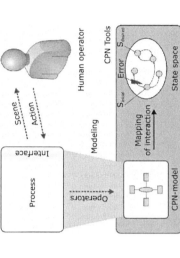

Example: Arcade game application

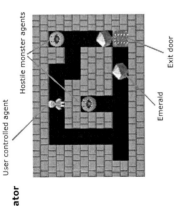

User controlled agent

Hostile monster agents

Emerald

Exit door

Goal of the human operator

- Picking up a certain number of emeralds
- Entering the exit door

Advantages

- Simple handling
- Custom levels and elements
- Replaceable by other technical processes

Gamrad, Oberheid, Söffker: Simulation of Human-Machine-Interaction for the Detection of Human Errors
© for all figures/illustrations by SRS U DuE

Modeling of interactions between human and game

Up

Left Situation Down

Right

1'(3,0,true,7,0,"Down",3,2,...

SITUATION

Actions of the agent

Situations used for modeling scenes

- Represented by one place containing one composite token
- Token is a set of data types representing the characteristics

Operators used for modeling actions

- Represented by transitions
- Related to one place
- Detailed on subpages

Gamrad, Oberheid, Söffker: Simulation of Human-Machine-Interaction for the Detection of Human Errors
© for all figures/illustrations by SRS U DuE

Error detection scheme

Synchronization from reality to model

- All performed operators as input
- Identification of the current node in the state space

Determination of goals

- Calculates all possible goals
- Uses observed user operators
- Filters specific user goals

Gamrad, Oberheid, Söffker: Simulation of Human-Machine-Interaction for the Detection of Human Errors
© for all figures/illustrations by SRS U DuE

Detection of the human error rigidity

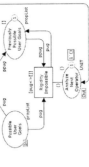

Definition of rigidity

- First condition

  The operator O_e is not directed to a goal situation.

- Second condition

  The operator O_e was directed to a previous goal situation.

O_e: Erroneous operator performed by the user

Applicable to other human errors classified by Dörner

Gamrad, Oberheid, Söffker: Simulation of Human-Machine-Interaction for the Detection of Human Errors
© for all figures/illustrations by SRS U DuE

Summary and future work

Summary

- Simulation of SOM using software tools for high-level Petri Nets
- State-space-based detection of human errors according to Dörner
- Formalization with SOM and related implementation with CPN Tools
- Realization of automated detection of the human error rigidity

Future work

- Detection of other kinds of human errors
- Analysis of interactions in more complex systems
- Processing of interactions in systems resulting in a large state space

Gamrad, Oberheid, Söffker: Simulation of Human-Machine-Interaction for the Detection of Human Errors
© for all figures/illustrations by SRS U DuE

17

Example: Simulation

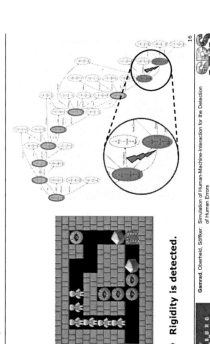

⬆ **Rigidity is detected.**

Gamrad, Oberheid, Söffker: Simulation of Human-Machine-Interaction for the Detection of Human Errors
© for all figures/illustrations by SRS U DuE

16

Multi-Modal Human-Machine Interaction and Cooperation in 4D Arrival Management

Hendrik Oberheid[1] and Dirk Söffker[2]

[1]Institute of Flight Guidance, German Aerospace Center (DLR),
D-38114 Braunschweig,
[2]Chair of Dynamics and Control, University Duisburg-Essen,
D-47048 Duisburg

1 Introduction

The contribution is concerned with the topic of arrival managment in air traffic control. Potential future processes and mechanism are discussed for controlling arrival traffic to an airport in an optimal manner, questioning which assistance systems and procedures are necessary to achieve this task. Specifically, two different research questions regarding arrival managment are in the focus of this presentation:

First, from a human-machine interaction point of view, the future communication between air traffic controller on the ground and flightcrew on the airborne side is likely to feature more and more elements of datalink based message exchange (instead of the conventional voice-only communication). This will lead to a mix of voice-based and text-based communication [1]. Selected human-factors issues and concerns which result from this move towards multi-modal communication are considered in this talk.

Second, the new communication protocols and planning mechansim are expected to feature increasing cooperation between air and ground. Part of the planning made by systems on the ground, will be based explicitly on user (i.e. aircraft) inputs and preferences. However, if the new interaction protocols and planning mechanisms are not designed properly, the self-interested behavior of individual aircraft can also run counter the realization of the global system objective. The contribution discusses the formal modeling of potential new protocols and their careful analysis as a key element for validating that the design objective can actually be reached. A CPN model [2] of a cooperative arrival mangement process is introduced which was built in order to analyze agents' incentives with decision-theoretic methods

2 What is Arrival Management?

Arrival Management

Aim of the arrival management process is to provide an optimal (safe and efficient) scheduling, guidance and control of arrival traffic to an airport. Independent of the exact procedures applying at a specific airport (which can vary significantly depending on the airport, local regulations and geography), the following four generic subproblems will have to be solved by any variant of an arrival management process:

1. *Sequencing*, i.e. establishing a favorable arrival sequence (sequence of aircraft) according to a number of optimization criteria,
2. *Metering* that is to calculate for each individual aircraft the target times over (TTO) certain points (fixes) of the respective arrival route appropriate for the aircraft's position in the sequence,
3. *Trajectory generation*, which means finding an efficient and conflict free route for each aircraft from its current position to the runway threshold in accordance with the TTO, and also
4. *Clearance generation*, that is, to provide detailed instructions to the controller with regard to the specific clearances which should be given to an aircraft in order to lead it along the route as planned.

Currently, the level of automation and kind of assistance systems used to support the human air traffic controller in his task may vary widely depending on the size of the airport and its technical development. It ranges from virtually no automation (radar display with traffic state information only) to working positions with integrated planning systems AMANs (Arrival Management Systems). These may offer sophisticated support for sequencing and metering. AMAN assistance functions for automated trajectory genereation and clearance generation are a topic of ongoing research but are expected to become industrialized in the future.

3 What is Cooperative Arrival Management?

The future - towards cooperative Arrival Management

Future procedures and mechanisms for arrival management are likely to feature a much closer integration of the respective actors and planning systems in the air and on the ground as it is the case today [3].

On a technical level, the closer integration means that a direct coupling and data exchange will be established between the ground-based arrival management system AMAN and the Flight Management Systm (FMS) on board of the aircraft via a digital datalink. Such a coupling is necessary to make better use of the advanced navigation capabilities of modern aircraft's FMS than is currently the case. Fuel and noise efficient trajectories can be calculated through the FMS but have to be coordinated with, and integrated into, the ground based planning. Nowadays voice radio is still the standard means of communication between air traffic controllers and pilots, making it impossible to exchange greater amounts of (numerical) trajectory data, which is necessary to lead the aircraft along the most preferable route or profile.

From an air traffic control point of view, the most important change may be that the trajectory will in the future no longer be commanded unilaterally to the aircraft by the air traffic controller. Rather the trajectory will be the result of some kind of bidirectional negotiation between air and ground wherein the user (aircraft) preferences and performance data are explicitly taken into account. The negotiation will result in a trajectory contract for a 4D-trajectory. The 4D-contract is an agreement on which trajectory is going to be followed by the aircraft and at which spatial point the aircraft is going to be at a certain point in time (see interaction diagram, slide 4).

4   Dual-modality Human-Machine Interaction

Dual-modality communication - control & and information flow
Dual-modality: Datalink vs. voice characteristics

Within the current aircraft fleet flying today, equipage levels in terms of datalink equipment and flight mangement system (FMS) functionality vary widely. The current DLR project FAGI (Future Air-Ground Integration) investigates the guidance of 'equipped' and 'unequipped' aircraft according to different procedures but within the same airspace. This aims at beginning to use the above mentioned cooperation protocols between air and ground to better exploit capabilities of modern fully equipped aircraft while providing the traditional procedures for unequipped aircraft.

Closely related to different, coexisting procedures in FAGI for equipped/ unequipped aircraft is the requirement for human operators to act within a dualmodality environment with data link and voice communication [4]. The issue concerns the flight crews of equipped aircraft and even more so the controllers. Flight crews of equipped aircraft will e.g. negotiate and exchange trajectory information via the data link medium. In parallel they will maintain radio contact with the controller for other tasks (e.g. handovers, short term conflict resolution). Controllers will guide equipped aircraft mainly via data link clearances. They will at the same time control unequipped aircraft exclusively via radio communication. Only flight crews of unequipped aircraft will be experience minor changes compared to today, apart from the fact that there will be no party line information on the voice channel for the equipped aircraft.

Existing research has shown that voice radio and data link exhibit different communication characteristics with different strength and weaknesses [5,6,7]. This affects their suitability for certain tasks from a human factors point of view. On the one hand, data link messages appear more suitable to transmit greater amounts of numerical data. This is why they are preferable for the negotiation of trajectory data and overfly times of equipped aircraft in FAGI. Data link messages also often show higher message quality (less error prone, misunderstanding/read back errors). They offer the potential to parallelize communication with different aircraft from a controller point of view and handle multiple open message transaction at a time, which can be cognitively highly demanding however. Also congestion of radio channels may be alleviated by data link. This improvise the chances of giving radio clearances to unequipped aircraft with good timing and precision enabling better adherence to 4D-planning. On the other hand, preparation, transmission and response time for data link messages is generally higher [5] which has been discussed as a potential source of additional controller workload, Continuous switching between modalities adds an additional 'modality monitoring task' to the task of the operator, and, due to the different response times of data link and voice, causes a need to for setting different 'mental timers' [5].

In FAGI the usage of data link as a medium for communication between ground and equipped aircraft introduces the technical possibility of having direct communication between airborne and ground automation, that is between AMAN and FMS, potentially without the need of human involvement. Such direct interautomation communication can be curse or blessing from a human factors point of view. On the positive side it enables the repetitive exchange of routine data between air and ground without continuously interrupting human operators work flows. It may thus help to avoid additional workload. On the negative side there would be a potential for leaving the human operator out of the loop if significant information was exchanged between system and the human operator was bypassed. Slide 5 highlights the differences in information flow with regard to the guidance of unequipped and equipped aircraft and illustrates that in the case of unequipped aircraft all transmitted information will be exchanged between human operators due to the absence of data link, while equipped aircraft communication allows system to system transactions.

5 Cooperation and Competition

Arrival Management - today and in future
Cooperative Arrival Planning and Incentive Structures

It has in the last paragraphs been pointed out that in many ways the novel control procedures and mechanisms outlined above will move the system towards a more cooperative system. This follows the aim of reaching better coordination and usage of limited resources and thus a higher global system efficiency. However, if the new interaction protocols and planning mechanisms are not designed properly, the self-interested behavior of individual agents (aircraft) can also run counter the realization of the global system objective. To understand that point, it is important to consider that the arrival process to a highly frequented airport inherently also contains a strong competitive element. At least during traffic peaks, different aircraft and airlines are potentially competitors with regard to limited airspace, route and runway resources. A position gain in the sequence for one aircraft will most likely be achieved at the price of some negative effect for one or more other aircraft. No two aircraft can use the same airspace or route segment at the same time.

On slides 7-8, a number characteristics are opposed on the nature of arrival management today and in potential future environments. It can be shown [3] that some of these characteristics (especially the developments towards bidirectional negotiation, deterministic planning systems and improved situation awareness for aircraft) potentially enhance the options of individual actors to manipulate a planning process in their own interest. In order to prevent this, the compatibility between incentive structures set by the mechanism and procedures and the behavioral expectations as formulated by the designer has to be actively validated and ensured.

6 Modeling

CPN Model of Cooperative Sequence Planning
State Space Calculation vs. Simulation
SeqPlanning Page - Simulation and Analysis Approach

This section of the presentation discussed a CPN Model [8] developed to analyze the behavior of the cooperative sequence planning process as outlined above. The entire model consists of 16 pages arranged in 5 hierarchical layers. All 16 pages form a tree structure departing from the toplevel page named Seq-Planning. An overview of the structure of the model is depicted on slide 9.

A special simulation and analysis approach was developed for the model, which also impacts the model structure. To realize that approach, the pages are organized into three different modules (M1, M2, M3) as indicated on the slide. The approach includes alternating and iterative use of state space analysis techniques and simulation for the three different modules. The results of the simulation or analysis phase of one module serve as an initialization for the next phase, realizing a cyclic process.

As an outline, Module M1 represents variations of aircraft submitted arrival times, the generation of potential candidate sequences and the evaluation of these candidates. A state space based approach is most suitable for this task since it allows that only the constraints with regard to time changes and for sequence building are modeled in M1. The computation of the full set of possible combinations is solved through automated state space generation. Module M2 performs a selection of candidate sequences generated in M1. In order to do that it queries the state space of module M1 and compares the candidate sequences with each other. M2 is executed in simulation mode. Module M3 integrates the results from the iteration between M1 and M2 and is executed in state space mode.

The SeqPlanning page is depicted in slide 11. It features five main substitution transitions and two regular transitions.

The substitution transitions SequenceGeneration, SequenceEvaluation, Sequence-

Selection and the regular transition SequenceImplementation together implement the behavior of the automated sequence planning system. Together the
four transitions map a perceived traffic situation as planning input to an implemented sequence plan as an output of the planning system.

The substitution transition ArrivalEstimateVariation in the lower part of the
picture represents the behavior of aircraft. These aircraft may vary their submitted arrival time estimates at certain points of the arrival procedure. Note
that these variations are external to the planning system and represent changes
of the traffic situation that the planning system has to react to.

7   Simulation Results

Example Scenario and Analysis
Analysis of a Single Planning Cycle
Analysis over Consecutive Planning Cycles

The section presents some preliminary results achieved through simulations
and analysis of the above model. The results contain both output produced with
a first approach to examine one single planning cycle as well as with the second
approach realized on the ProgressionGraph page to examine the behavior of the
system over a number of consecutive planning cycles.

The purpose of the simulation is to examine how the arrival planning system
AMAN reacts to changes in ETAs (Earliest Times of Arrival) which are submitted by the aircraft, and how these changes of ETA affect the aircrafts' position in
the sequence planned by the AMAN, and the target time of arrival the sequence
planner calculates for each aircraft on the basis of the new input data. Slide 12
shows the initial state of the simulation. Slide 13 shows the planned target time
of arrival of aircraft 'A','B' and 'C' plotted against the submitted earliest time
of arrival of 'A' and 'B'. The TTAs actually simulated with the CPN models are
represented by the triangles, squares and circles for 'A', 'B' and 'C' respectively.
From the perspective of the individual aircraft the gradient of the surface can be
interpreted as an incentive to adjust its behavior (speed/estimates) in one way
or the other. From the perspective of the air traffic control (or the perspective
of the AMAN alternatively) the most upper surface of the three is generally the
most interesting, as it reveals the time when the last aircraft in the considered
sequence will be landed.

Fig. 12 shows a fragment of a graph describing the system behavior of the
sequence planner over two consecutive planning cycles in reaction to variations of
the ETAs and LTAs. The same initial state of the example sequence is assumed
as for the previous analysis. Each node in the tree represents a resulting arrival
sequence as a function of the current value and history of ETAs submitted by
the aircraft. The results in this picture show that the planning result depends
on the history and timing of sumbissions by each individual actor, which could
potentially offer opportunities for manipulating the system. In the presented

case, however, no aircraft was able to improve its own outcome by systematic behavior, but aircraft A could potentially harm aircraft B by strategic action.

8 Conclusions

This presentation discussed a number of issues about current developments in arrival management for air traffic control. First, a move towards more cooperative procedures between actors on air and ground (air traffic control, Arrival Mangement System (AMAN), flightcrew, Flight Managment System (FMS)) was described. These interactions will rely increasingly on datalink communication for data exchange, leading to an environment with mixed text-based and voice-based communication. This brings up new issues from a human factors point of view some of which were discuess. Second, despite the inherent chances of the new cooperative procedures, new chances for system manipulation might also arise if the mechanisms are not designed right and incentive compatibility is not ensured. An example of a model-based approach using a CPN model was introduced to analyze incentives set by different system variants of a planning sytem AMAN.

Literaturverzeichnis

1. Oberheid, H., Temme, M.M., Kuenz, A., Mollwitz, V., Helmke, H.: Fuel efficient and noise-reduced approach procedures using late merging of arrival routes. In: German Aerospace Congress 2008, Darmstadt (2008)
2. Jensen, K., Kristensen, L.M., Wells, L.: Coloured Petri Nets and CPN Tools for Modelling and Validation of Concurrent Systems. Software Tools for Technology Transfer (STTT) **9**(3-4) (2007) 213–254
3. Oberheid, H.O., Söffker, D.: Designing for cooperation - mechanisms and procedures for air-ground integrated arrivalmanagement, Montreal, Canada (2007)
4. Dunbar, M., McGann, A., Mackintosh, M.A., Lozito, S.: Re-examination of mixed media communication: The impact of voice, data link, and mixed air traffic control environments on the flight deck. Technical Report NASA Technical Memorandum - 2001-210919, NASA Ames Research Center (2001)
5. Prinzo, V.O.: Data-linked pilot reply time on controller workload and communication in a simulated terminal option. Technical Report U.S. Department of Transportation Federal Aviation Administration, DOT/FAA/AM-01/8 (2001)
6. Helleberg, J., Wickens, C.D.: Effects of data link modality on pilot attention and communication effectiveness. In: International Symposium on Aviation Psychology, Columbus, Ohio (2001)
7. Lancaster, J.A., Casali, J.G.: Investigating pilot performance using mixed-modality simulated datalink. Human Factors **50**(2) (2008) 183–193
8. Oberheid, H.O., Söffker, D.: Cooperative arrival management in air traffic control - a coloured petri net model of sequence planning. In: International Conference on Application and Theory of Petri Nets and Other Models of Concurrency - ATPN, Xi'An, China (2008)

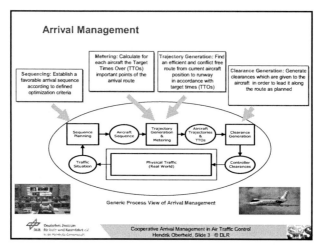

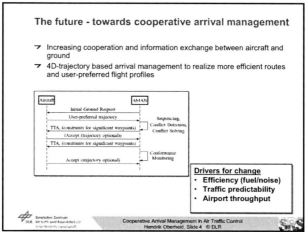

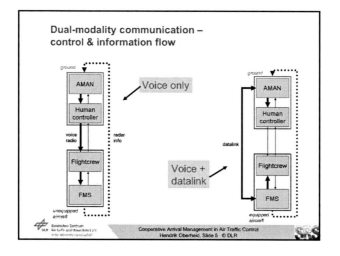

Dual-modality: Datalink vs. voice characteristics

Voice and Datalink exhibit different communication characteristics with strength and weaknesses

Datalink Pros:
- ➤ more suitable for large amounts of numerical data
- ➤ higher message quality
   - ➤ less error prone, misunderstanding/read back errors
- ➤ potential to parallelize communication with different aircraft
- ➤ potential to handle multiple open message transactions at a time

Datalink Cons:
- ➤ preparation, transmission and response time is generally higher
- ➤ Continuous switching between modalities adds an additional "modality monitoring task"
- ➤ causes a need to for setting different "mental timers"
- ➤ prevents „party-line" effects
- ➤ ...

Cooperative Arrival Management in Air Traffic Control
Hendrik Oberheid, Slide 6 © DLR

Arrival Management - today and in future

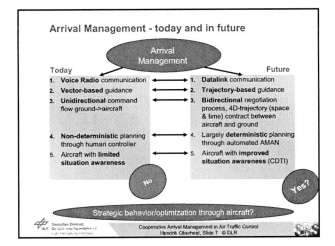

Arrival Management

Today
1. **Voice Radio** communication
2. **Vector-based** guidance
3. **Unidirectional** command flow ground->aircraft

4. **Non-deterministic** planning through human controller
5. Aircraft with **limited situation awareness**

Future
1. **Datalink** communication
2. **Trajectory-based** guidance
3. **Bidirectional** negotiation process, 4D-trajectory (space & time) contract between aircraft and ground
4. Largely **deterministic** planning through automated AMAN
5. Aircraft with **improved situation awareness** (CDTI)

No

Yes?

Strategic behavior/optimization through aircraft?

Cooperative Arrival Management in Air Traffic Control
Hendrik Oberheid, Slide 7 © DLR

Cooperative Arrival Planning and Incentive Structures

Facts:
- ➤ Bidirectional planning transfers degrees of freedom (DOF) from ground to aircraft
- ➤ Groundbased planning (AMAN-System) based on aircraft information according to fixed rules
- ➤ Aircraft gains situation awareness

   Strategic usage of DOF becomes possible in principle!

Questions:
- ➤ Can we assume cooperative behavior to ensure global system efficiency?
- ➤ Which incentives exist from local viewpoint of individual actor?
- ➤ What behavior do we expect?

Model based approach & decision theoretic analysis

Cooperative Arrival Management in Air Traffic Control
Hendrik Oberheid, Slide 8 © DLR

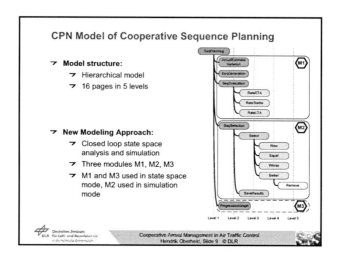

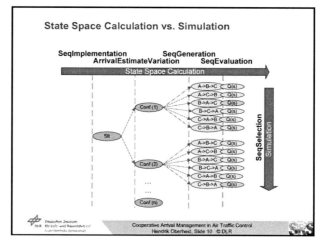

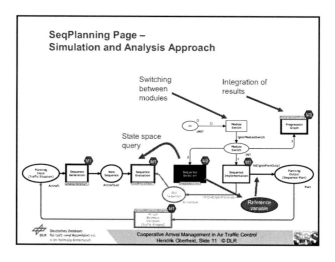

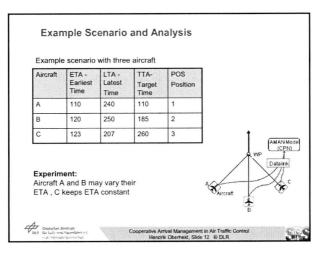

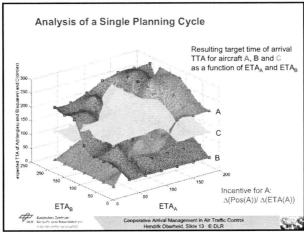

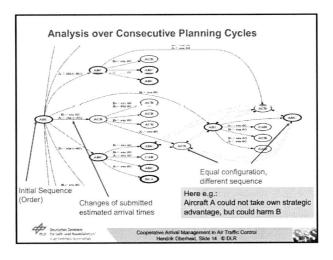

Trust and Reputation

Jens Hardings

Computer Science Department, Universidad Católica de Chile, Santiago

jens@hardings.cl

Abstract. The usage of information technologies (IT) can provide a wide range of services to political activities. Among the services are coordination of interest groups, administration of working groups and committees, management of registered voters and members of political parties, coordination of advocates and fundraising, among others. However, one of the most interesting issues regarding the use of IT in politics is still a promise: using IT to make the interaction between citizens and elected representatives scalable.

Due to human limits, it is not conceivable that every person is able to represent her interests in all of the topics relevant to her. This has two important consequences in the way we organize a democratic society. First, we cannot possibly have a direct democracy, where every citizen is responsible for the outcomes of all political decisions. The global interconnected interaction is too complex, and we could not possibly address all of the issues that are relevant to every person. Second, the representatives cannot be elected based on all of the relevant topics for every voter. Only the most significant topics for the majority of the population will be present in discussions such as debates and government programs. This problem has been labeled as "The Long Tail" in the retail business, and it has been shown that using IT to improve access to less massive products can create a whole new and lucrative business. The same way, politics can benefit of paying attention to real and meaningful problems for many people, problems that are currently ignored or left in hands of few decision makers who are not necessarily representative.

The goal of using IT in politics is to enable a better democracy, to improve current situations and avoid opening the door to new problems that play against that goal. Current situations that make the system unfair is that the more resources a person has, the more she can influence the political system to follow her agenda. The same can and does happen in systems that are based on IT if we take no precautions to avoid it. Additionally, if we limit the participation of people in the political arena to persons with Internet access, we are creating a new barrier to the participation. We have to make sure that the access to participation does not get worse by introducing changes, and specially in a country like Chile, with very pronounced income inequalities, this is a very sensible area.

We propose to create a medium-scale system that helps us to understand the consequences and propose the best solution for such initiatives. The first target is the municipal government, in which the citizens have a more direct relationship with the elected authorities, and the traditional media do not offer much help in the communication among citizens and their representatives. The system will allow people to pose their suggestions, complains and doubts in specific areas, where they will first be managed by their peers. The relevant topics and opinions will then stand out and be addressed by the elected representatives and the municipal workers. By using this system, the objective is to enable citizens to present their concerns and, when they have merit, enable them to be used as part of future policies.

A secondary objective is to give the citizens the tools necessary to assess the work of their elected representatives. When their concerns are taken care of efficiently, they would want them to be reelected. However, when there is evidence that the administration is not as good as it could have been, or that the elected representatives prefer to benefit some minorities over the bulk of the population, they would express their rejection in the next election using objective data.

1 Introduction

The usage of information technologies (IT) can provide a wide range of services to political activities. Among the services are coordination of interest groups, administration of working groups and committees, management of registered voters and members of political parties, among others. We can see a huge interest in using IT for voting processes, with varying degrees of success, and specially acceptance, because of privacy, accuracy and accountability issues (Mercuri, 2002). Since the Howard Dean campaign in the USA 2004 presidential election, there has been a strong evidence of the utility posed by IT in other processes: aligning advocates and raising funds (Shapiro, 2003), confirmed again in the 2008 primaries. However, one of the most interesting issues regarding the use of IT in politics is still a promise: using IT to make the interaction between citizens and elected representatives scalable and bidirectional.

Due to human limits, it is not conceivable that any person is able to represent her interests in all of the topics relevant to her. This has two important consequences in the way we organize a democratic society. First, we cannot possibly have a direct democracy, where every citizen is directly responsible for the outcomes of all political decisions. The interaction with all other people is too complex, and we do not have the capacity to address all of the issues that are relevant to every person. Second, the representatives cannot be elected based on all of the relevant topics for every voter. Only the most significant topics for the majority of the population will be present in discussions such as debates and government programs. This problem has been labeled as "The Long Tail" in the retail business (Anderson, 2006), and it has been shown that using IT to improve access to less massive products can create a whole new and lucrative business. The same way, politics can benefit of paying attention to real and meaningful problems for many people, problems that are currently ignored or left in hands of few decision makers who are not necessarily representative.

The promise of IT in the political arena is to allow every citizen to express their will regarding every issue that is relevant to them. This is not a new issue since it has been around for over ten years (Grossman, 1996; Hague, 1999; Wilhelm, 2000) without much impact. Notable exceptions are the organization of the crowds that finally ended the scandal-ridden presidency of Joseph Estrada in Philippines, in which the people coordinated efforts using cell phones and their short messaging system to avoid police anticipation (Rheingold, 2002). Other examples are the resignation of senator Trent Lott in the USA, after activists dug up his racist comments, and the turnover of the presidential elections in spain in 2004 after the terrorist attacks were wrongly attributed by the government to ETA. However, these are still just influences in how to elect and supervise the representatives in more efficient ways, and do not represent a fundamental change in the way politics is made.

If we were able to vote from the comfort of our homes, and the results can be available almost instantly after receiving every electronic ballot, the possibility of voting on every single decision would be tempting. However, such a mechanism is not scalable and has one big limitation: the mechanism to define what is to be voted, and what the options are, has to be defined somehow, leaving us with the same initial problem. Therefore, it is necessary to use IT in a way that every person could raise her own issues, and this way treat every relevant topic, not just the ones massive enough. Currently, in order to raise an issue a person needs huge resources in order to mobilize media and get to enough citizens and decision makers in order to just make their case known.

The goal of using IT in politics is to enable a better democracy, to improve current situations and avoid opening the door to new problems that play against that goal. Current situations that make the system unfair is that the more resources a person has, the more she can influence the political system to follow her agenda. The same can and does happen in systems that are based on IT if we take no precautions to avoid it.

Additionally, if we limit the participation of people in the political arena to persons with access to, e.g., the Internet, we are creating a new barrier to the participation that did not exist previously. We have to make sure that the access to participation does not get worse by introducing changes, and specially in a country like Chile, with very pronounced income inequalities, this is a very sensible area.

We propose to create a medium-scale system that helps us understand the consequences and propose the best solution for such initiatives. The target is the municipal government, in which the citizens have a more direct relationship with the elected authorities, and the traditional media do not offer much help in the communication among citizens and their representatives. Once we have begun to understand the impact of such a system in the medium-scale application, we can start to adapt it to a larger scale. The system will allow people to pose their suggestions, complains and doubts in specific areas, where they will first be managed by their peers. The relevant topics and opinions will then stand out and be addressed by the elected representatives and the municipal workers. By using this system, the objective is to enable citizens to present their concerns and, when they show to have merit, use the ideas and suggestions and implement them. This principle has been termed "Emergent Democracy" (Ito and Lebkowsky, 2004), and we can find examples in which crowds tend to act smarter than individuals (Surowiecki, 2005).

A secondary objective is to give citizens the necessary tools to assess the work of their elected representatives. When their concerns are taken care of efficiently, they would want them to be reelected. However, when there is evidence that the administration is not as good as it could have been, or that the elected representatives prefer to benefit some minorities over the bulk of the population, they would express their rejection in the next election using objective data.

2 Requirements

When proposing information systems to solve the problems of making direct involvement of people in every political decision of interest to them, we need to address several requirements.

Trust towards the underlying system When people do not trust the system, they will not be compelled to participate. Democratic systems based on voting where every individual may participate together with trusted parties to overview the voting process. While no political party trusts the other participants, the system provides enough transparency to trust the process of a political election. Also, the system has to be robust to resist abuses. This is detailed further below.

Wide Scale adoption This presents two different challenges. One is to find a system that may work on a large scale, given the restrictions like transparency and others.

The scale being targeted is at least nationwide, but ideally such a system should work considering the whole world population. The second challenge is to get people to participate, because the system will only work as it gains critical mass.

Robustness Strongly related to trust, the robustness of such a system when subject to attacks is of vital importance. If we succeed producing a useful information system that helps us decide our future in politics, this system will automatically become the number one target of several attacks intending to influence those decisions. It is necessary for the system to be designed to avoid these attacks to be effective, or at least being able to detect and act upon attacks.

Do no harm It is a reality that not every person has equal access to online communities, and thus to any proposed information system on a world-wide scale. The digital divide is a serious barrier, and if we limit citizen participation to those who already have access, the situation would be worse than currently for the ones who have no access. It is necessary to avoid this situation and improve the current opportunities for every participant.

Avoid concentration of power When a small group has excessive power or influence on the rest of the population, the potential threat of misusing that power becomes more evident. Nowadays the concentration of power takes the form of wealth, control of mass media and influence on decision makers. In an information system based on reputation, one form of power concentration could be an overly high reputation for some users, who might use it in their own benefit. It is expected that the design of the system avoids this kind of situations, even in presence of active efforts on behalf of several participants (see Robustness).

Considering the presented issues, it becomes clear that this project needs to be considered as a multidisciplinary activity. Therefore, it is important to define the requirements based on the socio-cultural, political and technological aspects. While this project is focused on the technological part, it considers work and collaboration with other disciplines, namely Sociology, Political Science and Communication.

Current information systems, specifically those designed to be adopted on a wide scale, are generally designed to be able to interact through a web interface. This interface might not be the only one, but a system of the scale to be considered for this project certainly needs the kind of access provided by a web interface. A current trend in web applications has been framed as social networks or "web 2.0" by O'Reilly (O'Reilly, 2005), and might present a set of criteria that promotes wide scale collaboration of users as the examples suggest. It is necessary to assess whether those characteristics are important to be implemented in the proposed information system.

3 Reputation Systems

Reputation systems have been used in several contexts, the most prominent being auction sites and service provision (Resnick et al., 2000; Jøsang et al., 2007) in which the reputation is formally calculated according to a model and the implicit or explicit input to the system. In auction sites feedback is generally only possible once for each transaction, where the buyer rates the seller and vice versa. This tends to legitimate each vote. In other cases, the reputation is not calculated using an internal model but rather measured externally, as in the case of Technorati to measure weblog popularity.

Zhang et al. propose five basic trust parameters and a computational trust model to measure the trustworthiness of participants in online auctions (Zhang et al., 2005). The parameters include feedback rating, the trust value of the last period, reliability of raters, decay of feedback rating and value of a transaction. While the concept of reputation has been used in many fields such as economy, sociology, and computer science, the literature has only recently begun emphasizing the advantages of using reputation systems for the establishment of trust relationships in large scale networks (Boukerche et al., 2004; Buchegger and Boudec, 2004; Cahill et al., 2003; Keane, 2006; Pirzada et al., 2004).

In general, the reputation is meaningful only inside the particular system or model in which it is calculated. This allows variation of parameters used in the calculation that can be used to avoid abuses and adapt the system to changing conditions and improve the model constantly. For example, it is not interesting to be able to identify the real person behind a particular Weblog, or to know who the person is that is selling you an item through an auction site. What matters is that the system ensures that we are talking to the same entity that has earned a particular reputation. This reputation can also be influenced by analysis such as verifying consistency among discourse and action.

While large scale reputation systems do exist, it is not obvious whether they might be implemented considering the specific requirements presented for this project. Let us consider the Google Pagerank system Page et al. (1998). It is a reputation system where each web page available on the internet has a relevance (reputation) assigned to it and can give relevance to other pages by providing a link. More relevant pages give more relevance to their linked pages, and the new values get calculated periodically based on the current values. For this system to work, Google needs to adjust certain parameters which are kept secret and this way potential abusers cannot optimize their tactics. However, in the proposed system we need the system to work on a large scale but presenting all of the information for transparency. It is not clear whether the system can work at that scale, be robust and at the same time deliver all information to every interested party.

Several existing systems have been found vulnerable to either specific or general attacks. This is the case with the eBay feedback system, General/Situational Trust (Marsh, 1994), Multidimensional Trust (Griffiths, 2005), Tran and Cohen (Tran and Cohen, 2002, 2004), Sporas/Histos (Zacharia, 1999), Regret (Sabater and Sierra, 2001), as well as Yu and Singh (Yu and Singh, 2002). Common problems include change of identities (Zhang and Fang, 2007), inadequate protection of privacy of references (Resnick and Zeckhauser, 2002) and dishonest feedback (Tuan, 2006; Whitby et al., 2005). According to Pérez et al., users are demanding capabilities that are not available in present systems (Pérez et al., 2006). The authors propose a set of principles that should be considered in any digital democracy platform:

- Freedom of speech, whereby all users of the platform can express themselves with no fear of reprisals in the present or in the future.
- Equality, whereby the opinions of all citizens carry the same importance.
- Mutual respect. Opinions expressed publicly must observe certain rules that have been defined and accepted by the participants in the forum themselves.
- Determinate duration of discussions. Subjects for discussion shall have a lifetime that is agreed and known by users when the debate commences.

- Citizens should have robust probes in order to verify that the system is functioning properly. That means the system must be auditable.
- Validation of conclusions either by consensus or through a vote. In the latter case, the system must ensure a clean voting process.

3.1 Components and actors

A reputation system comprises several components and actors. We can identify:

Rater Collector who gathers ratings from agents

Processor Processes and aggregates the information

Algorithm Used by the processor to calculate an aggregated representation of an agent's reputation. It is the metric of the reputation system

Emitter Makes the results available to other requesting agents

The participants can be classified according to their behaviour in:

Honest behaves honestly, rates correctly, should be rated as trustworthy

Malicious initiates good, neutral and bad transactions by chance. Tries to undermine the system with his rating behavior and rates every transaction negative

Evil or Conspirational group of agents knowing each other, rating each other positively

Selfish blocks all inquiries by other agents and refuses to rate his transaction partners

Disturbing tries to build a high reputation, such that the other agents trust him, with making good transactions and correct rating. Then he switches to a malicious behavior until his reputation gets too bad and then starts from the beginning

4 Trusting the System

One important aspect regarding the system is that people who use it can trust the results. This is best done by enabling transparency in addition to interactivity (Welch and Hinnant, 2003), which is already being provided by this system. Transparency can be provided in two fields, providing access to data on one side, and the algorithms and their implementation on the other. The former allows data to be available for any person to perform her own calculations and verifying the results. This is the equivalent of providing the detailed results for each ballot box in an election, so that each political party can keep track of each individual result and thus calculate the global outcome of said election, verifying that there is no attempt to commit fraud.

On the other side, the calculations themselves are performed by the software that runs the system. Unlike vote counting, the algorithms are not only unknown, but constantly changing. In this case, the only way to truly provide transparency is to provide not only the data, but also the source code that is used to process and manipulate the data. Thus, the system has to be necessarily available under an open source license, which allows any interested party to verify the results, and even manipulate the algorithms to try out variations and find out how possible differences came to be. As both the software and the data are in constant evolution, it is not enough that they are available to download. In order to verify the transparency, it is necessary to provide each relevant step in that evolution. For the software, each modification has to be available, as well as the information of when each modification was activated on the main system, so that it is possible to recreate the system from its initial to the current state. Additionally, each input and output the system received from its initial to its current state has to be available. So, it is necessary to provide the initial data, the input to the system that have an impact on the data and the final state. With this information, the objective to provide a repeatable sequence of events that lead from one state to another, providing full transparency on the inner workings of the system, is finally fulfilled by publishing both the software and data, the latter including the input to the system, in periodical intervals.

As a side effect, it would be possible for any person or institution to provide exactly the same system from one. However, the authentication needs to either be kept private or use public key cryptography. It can be expected that users will collaborate more widely when they control their privacy level and can choose whether they want to publish their contributions in a similar way it influences the participation in surveys (Singer et al., 1993). Thus, it is necessary to decide whether the data available to any person to download and use will include the specific information of the user accounts, which could contain private information. Obviously the users have to accept the privacy policies upon registering, so this decision needs to be taken, although it can be different for every implemented system, before any data is gathered. In the case the data does not include private information, it is necessary to issue a transformation that allows the users' actions to be represented in some way, without necessarily linking them to their personal information.

In order to assure that the published data is not changed afterwards, the datasets and software releases will be digitally signed. This avoids the original site to modify the dataset afterwards, increasing the trust people will assign to this site. It also avoids anybody to claim having downloaded a dataset that was never published as such.

One of the hypotheses in this project is that people will participate actively when they can trust that their work is being considered. When people participate actively in a community where they see all of their suggestions are being ignored, or they do not receive any feedback on why their ideas did not succeed as they suggested and were thus left behind, they will promptly stop their active participation. In this project, all of the ideas presented by any interested person will have a chance to be evaluated by peers, experts and representatives of the community. This way, it should be possible to have an explanation on why a particular suggestion was not taken into account, rather than just vanish among the huge amount of interactions between community and municipality, where it is not feasible to give feedback to every citizen proposal. This should be an incentive to people for improving their participation rather than stopping it upon difficulties.

5 Participation and Digital Divide

Participation is critical for a system like this, since it depends on achieving a critical mass of users that express their opinions and suggestions. It is necessary to show the users the benefits of voluntary participation, and also to have some initial data to get the system working as expected as soon as possible. When future users can start by watching the process, verify that it actually works and then begin to participate actively, the entry barriers will be much lower.

Additionally, one of the propositions of a system like this is to improve the democratic system in some way. This means to provide better opportunities for everybody to participate, and their opinions or suggestions to be heard when they have merit. But when we limit this system to participation among people that have access to the Internet, this would also restrict the opportunities and worsen the situation rather than improving it (Zuckerman, 2004).

While a 76,4% of the 10% highest income group in Chile has Internet access, the number falls to 27,8% (3 years ago it was only 16,2%) for the 10% lowest income group in 2006 (Mideplan, 2006). This reflects a problem that is difficult to overcome, and it has to be considered when evaluating the representativity of a system that needs internet access in order to participate. As this project aims to improve the current situation and allow a fair access to every interested citizen in participating in the relevant topics, the access to the information has to be addressed. Therefore, this project considers the preparation of several instances that promote active participation using several channels. One of the most suitable system, that is also scalable and replicable, is the preparation of access to participate using the proposed system. The needed infrastructure can be provided through several possible ways, using currently existing institutions. These include Infocentros, nationwide coordinated internet centres where newcomers can get help to start using the network, as well as neighbours organizations (juntas de vecinos), sport centers and the local government (municipalities) infrastructure.

However, the infrastructure alone might not be enough to assure active participation, initial trust towards the system and specially the basic knowledge to use it. It is necessary to provide some assistance to ignite the active use of the system, specially among people that do not use Internet technology on a daily base. The effort on behalf of this project is to prepare the instructors that will provide support to the end users. They need to understand the system, how it works and what is expected on behalf of the users so that their participation is as long-term as possible. It is necessary that the users understand that they can use the system not only to send suggestions and reports, but also to obtain feedback on what happens with their participation.

Several projects in Chile indicate that, when certain conditions are met, the participation of capable people is high. We can identify the some examples that share their strong IT-based roots in their organization, including Atina Chile, a political initiative that started around a successful web community that even had active involvement in the creation of a new political party, Chile Primero. Citizen Newspapers, most of them related to the previously mentioned initiatives, have shown an interesting success and participation rate. Educalibre is a community interested in improving the use of open source software in the educational arena, and has gathered a heterogeneous group of people interested in education, including teachers, students, programmers and others. Promesas.cl was a simple web site dedicated to harvesting political promises in order to later evaluate the presidential candidates. Nuestro PC is a spontaneous initiative for a counteroffer to a government-backed announcement of a particular offer for an entry-level computer available for low-income families. Liberación Digital is a citizen initiative that strongly rejected an alliance among the government and a software provider, getting access to several instances where the national digital strategy was defined. LiberaciónSI is a political initiative proposing a network neutrality bill, and the Library of the National Congress (Biblioteca del Congreso Nacional, BCN) has been improving access to bill dicussion, including activities in social networks such as Facebook. These projects are also a starting point to offer the use of the system to be developed for this project, making use of the experience, enthusiasm and technical knowledge of the people participating in these groups. As they already are used to several of the proposed characteristics, they can provide the initial data that will get the systems going. Once the systems have started their operation and the first results can be seen, it is easier to attract new users, since they can already see the whole process. We maintain the idea that reputation systems facilitate bridging the on-line world and the real-world field, leading to real political action from the discussion and vice-versa.

6 Security

As any other system where strong personal interests can conflict, there will be incentives to commit fraud that would benefit some of the participants. This is particularly true when considering a system that has distributed reputation management (Yu and Singh, 2003) and when, as in the proposed project, the outcome of the system would benefit or damage assets outside of the system. As a matter of fact, we are poorly prepared for controlling markets and communities where the participants' behavior is governed by self interest, or even worse, by a combination of selfish, malicious and criminal intentions (Jøsang and Haller, 2007). In this project, the proposed system is expected to gain a respectable level of trust so that the assessment of the elected representatives would imply gain or loss of votes in the next elections.

Therefore, one of the fundamental tasks early on in the implementation is to detect and avoid the possible mechanisms that could be used to commit fraud. Digital signatures have already been discussed as a way to provide not only transparency, but tamper-proof transparency, so that nobody can undetectably modify data or software after it has been released. The system has to be robust enough to avoid the manipulation of results on behalf of a reduced set of interests.

It is to be noted that, as the system does not necessarily map all users to real people, one particular person or institution can theoretically create a huge number of users inside the system. It is thus possible for one person to create a great number of users and this way try to amplify his votes. Countermeasures can be implemented so that in these cases, the votes are not amplified in a linear way, by penalizing the reputation of the users that behave this way.

7 Application Area

The system proposed in this project would allow citizens to participate directly in a community that constantly evaluates every contribution and emphasizes the important issues and those that represent the agreement of a majority of the participants. To do so, the system will make use of several characteristics of what has been identified as "Web 2.0" The most important characteristics are the ones related to social software, that is, software that enables people to interact, form online communities and collaborate using computer-mediated communication.

A municipality manages a wide range of resources, and the citizens living in their jurisdiction generally include experts in every one of the aspects inside that range.

Thus, the "customers" of the municipality almost always include experts that are as qualified as the people in charge of the particular issues. This is a perfect example of what von Hippel states in "Democratizing Innovation" (von Hippel, 2005), in that the customers are the most capable of suggesting innovations to their providers, and the latter should work together with their customers to seize that opportunity.

The system would be applied in several areas in which each municipality sees fit. Urban Planning (Urbanismo): people know best about specific areas where it is best to build a new park, what infrastructure their children prefer in playgrounds, what kind of plants grow best in a specific condition. These suggestions and reports of specific problems (a broken bench, missing lights, etc.) will be available for the responsible person to take action, and the results will be evident to any interested person. Security (Seguridad Ciudadana): the participation of citizens in this area can be classified into two categories: incident reports and expert opinion. Although it is not possible to identify whether a person is an expert on the topic, the opinions can be ratified by some previously known expert to evaluate its accuracy. Security issues can also benefit from reports of trusted sources, such as the police complaints filed for a certain period. Traffic (Tránsito): the reports in this area will represent suggestions, incidents such as excessive congestions, malfunctioning traffic lights or other signs as well as non-functional design of intersections or other traffic infrastructure.

8 Expected Results

A project like this is expected to produce several results. The most important result is the knowledge and experience in managing a medium-scale public participation system. This experience will produce several documents detailing the contributions, as well as publications in peer reviewed journals. Concrete results include:

- software and instructions necessary to deploy medium scale systems on a local government level
- a set of actually implemented and future improvements to the reputation system being proposed
- a document detailing the results of testing and validation performed in pilot projects
- a set of best practices based on the experience deploying the pilot projects
- a detailed plan to implement and deploy a large-scale system based on the medium scale pilot projects
- expressions of interest on behalf of public institutions to coordinate joint project for a large scale public participation system

9 Methodology

A starting point is local government, on the municipality level. The peculiarities of this project require a multidisciplinary approach, so an important part of the effort in this project is devoted to generate and fortify the necessary connections that will allow us to approach this problem and provide successful solutions.

Also, the funding will be targeted at financing students working in several disciplines as needed. While the central research focus is the usage of Information Technologies, several activities will require work on specific issues related to Sociology, Communications and Political Science, with outcomes that will be used as input to the IT-specific research.

9.1 Determining specific technical requirements and available reputation systems

We need to determine the specific technical requirements for the application of systems in the scope of this project. This requires an important input: the socio-cultural, communications and political science requirements for such a system to have the chance of becoming successful. Both results will be obtained by students (tesistas) and assistants (ayudantes) in the relevant disciplines, guided by the principal investigator and in connection with people directly involved in local government. In this phase we expect to argue on the convenience of explicit ratings versus implicit behavior or anonymity versus identification, user-specific issues, among other characteristics available in reputation systems.

Once the requirements are defined on the various aspects, we need to focus on how the currently available reputation systems and strategies can keep up. The criteria is to determine the applicability, strengths and weaknesses of each reputation system and strategy when faced with the various technical requirements.

As the systems become mature enough, it will be possible to have stable models and specifications that would make the isolation between each system less necessary from the technical point of view. We can already find specifications like OpenID where any provider can verify an identity by simply adhering to the defined standard. However, the reputation is an indicator of how much trust an entity will deserve, so the choice is not technical at all. If one system will let another one modify the parameters on which it calculates the trust assigned to a user, it implies trusting that other system.

In the case of the municipalities, the intention is to have a mapping of at least some of the modeled entities to people and institutions in the "real world", that is, identify the responsible people behind certain actions, and being able to link those actions with the according interventions inside the system. If we do not provide this kind of mapping, the system would work without many problems. When a problem is presented, gathers enough relevance to be one of the priorities and finally gets resolved, it is possible to infer these situations by just using the input of peers, without giving a special meaning to any participant. Also, the representatives can be evaluated based on their performance to act on the proposed problems without that feature.

However, we can enrich the model by considering additional aspects known about specific participants. A report issued by a trusted entity may be weighted differently than a relatively anonymous user. It is easier to discover a cheat when the sources are well known, and when it is discovered, some action can be taken. Thus the value of the information with known source is higher and should be rated as such. This also helps when setting up the initial values for reputations, while the critical mass is being reached. Certain well-identified parties will have a higher initial reputation and can influence the system. While this gives them power to abuse the system, the fact that they are identified and thus liable for their actions may have an effect. Any misbehavior can be detected and have consequences for the responsible, moreover when these consequences are not bound to the particular system.

9.2 Implementation of the defined information system

Based on the small-scale prototype developed in previous works, the information system will be made more usable addressing efficiency, user interface design and other relevant aspects for the user experience, as well as compatibility with relevant information sources. Using agile software development techniques, prototypes will be available early on, allowing field tests that will allow to correct errors and propose new ideas before investing much time in following a less ideal path. This early testing will also allow us to evaluate the performance in medium-scaled environments and act as necessary.

9.3 Improvement of results and extraction of data

The results of the processes can be continually improved. However, in this phase we will apply several strategies that aim at improving the results substantially. Some of the strategies include aspects such as:

— Usage of data-mining algorithms to extract information that will enable the system to improve the results, such as clustering of participants into groups based on common interests or other criteria
— Definition of taxonomies for particular problems, which allows the system to identify context based on vocabulary used by users. Intend to perform this automatically, updating the resulting folksonomy.

9.4 Analysis and improvement of robustness

The robustness when facing attacks is a central point in this project. This work will be performed by a student (tesista) under the guidance of the principal researcher who will take the proposed systems, identify attacks that can be successful and propose countermeasures to avoid those attacks. It has to be noted that the attacks are not limited to technical attacks, but also to the socio-cultural and political possibilities. In this regard, it is similar to research on e-voting.

9.5 Deployment plan

The final results include a deployment plan. To generate this plan, it is expected that the technological difficulties will be minor, so this part will focus on political and communicational as well as social strategies that can improve the odds of success when deploying the studied information systems either as a pilot plan or on a wide scale. This work will thus depend on the collaboration with the different disciplines that have been mentioned, as well as the experience gathered during the pilot implementations with local governments. The focus of this task is to develop a document that supports deployment in local and nation-wide government, as well as usage on behalf of national and international NGOs.

9.6 Determine the applicability of the results to other processes

The resulting information system can be considered as a generic platform that is used within the specific context of a public participation framework. However, improvements made to the system might be useful in other contexts, such as electronic voting, enterprise management, large scale resource planning or online child safety, among others. Part of this work is to establish possible externalities by applying algorithms or methodologies as well as the software in these and other areas.

9.7 Project evaluation

The last phase includes a practical evaluation of the project, as well as future tasks to be carried out in order to improve the developed systems.

9.8 Municipality Coordination

One important point that needs to be pursued during the development of the whole project is the coordination with other institutions that can benefit from the results of the project. The objective is to collaborate on several fronts:
− Implementation of pilot projects
− Provide access to students for determining requirements (part 1 of this section)
− Feedback on needs and experiences
− Explore possible application areas that benefit Municipalities, ordered by impact
− Training of monitors who will assist people using the information systems inside municipalities

10 References

1) Anderson, C. (2006). The Long Tail: Why the Future of Business is Selling Less of More. Hyperion.
2) Boukerche, A., El-Khatib, K., Xu, L., and Korba, L. (2004). Anonymity enabling scheme for wireless ad hoc networks. In Global Telecommunications Conference Workshops, 2004. Globe ComWorkshops 2004. IEEE, pages 136–140.
3) Buchegger, S. and Boudec, J. L. (2004). Nodes bearing grudges: Towards routing security, fairness, and robustness in mobile ad hoc networks. Proc. of the 10th Euromicro Work. on Parallel, Distributed and Network-based Processing, pages 136–140.
4) Cahill, V., Gray, E., Seigneur, J.-M., Jensen, C., Chen, Y., Shand, B., Dimmock, N., Twigg, A., Bacon, J., English, C., Wagealla, W., Terzis, S., Nixon, P., Serugendo, G. D. M., Bryce, C., Carbone, M., Krukow, K., and Nielson, M. (2003). Using trust for secure collaboration in uncertain environments. Pervasive Computing, IEEE, 2:52–61.
5) Griffiths, N. (2005). Task delegation using experience-based multi-dimensional trust. In Proceedings of the fourth international joint conference on Autonomous agents and multi-agent systems, pages 489–496, The Netherlands. ACM.
6) Grossman, L. K. (1996). Electronic Republic: Reshaping American Democracy for the Information Age. Penguin (Non-Classics).
7) Hague, B. N. (1999). Digital Democracy: Discourse and Decision Making in the Information Age. Routledge, 1 edition.

8) Ito, J. and Lebkowsky, J. (2004). Weblogs and emergent democracy http://joi.ito.com/static/emergentdemocracy.html. Retrieved May 6, 2008.

9) Jøsang, A. and Haller, J. (2007). Dirichlet reputation systems. In Availability, Reliability and Security, 2007. ARES 2007. The Second International Conference on, pages 112–119.

10) Jøsang, A., Ismail, R., and Boyd, C. (2007). A survey of trust and reputation systems for online service provision. Decis. Support Syst., 43:618–644.

11) Keane, J. (2006). Trust based dynamic source routing in mobile ad-hoc networks. Master's thesis, Trinity College Dublin. Department of Computer Science.

12) Marsh, S. (1994). Formalising Trust as a Computational Concept. PhD thesis, University of Stirling.

13) Mercuri, R. (2002). A better ballot box? Spectrum, IEEE, 39:46–50.

14) Mideplan (2006). Encuesta casen 2006. MIDEPLAN, Division Social, Gobierno de Chile.

15) O'Reilly, T. (2005). What is web 2.0: Design patterns and business models for the next generation of software. http://www.oreillynet.com/pub/a/oreilly/tim/news/2005/09/30/what-is-web-20.html. Retrieved May 6, 2008.

16) Page, L., Brin, S., Motwani, R., and Winograd, T. (1998). The pagerank citation ranking: Bringing order to the web.

17) Pérez, E., Gómez, A., Sánchez, S., Carracedo, J. D., Carracedo, J., González, C., and Moreno, J. (2006). Design of an advanced platform for citizen participation committed to ensuring freedom of speech. J. Theor. Appl. Electron. Commer. Res., 1:58–71.

18) Pirzada, A., Datta, A., and McDonald, C. (2004). Propagating trust in ad-hoc networks for reliable

19) Routing. In Wireless Ad-Hoc Networks, 2004 International Workshop on, pages 58–62.

20) Resnick, P., Kuwabara, K., Zeckhauser, R., and Friedman, E. (2000). Reputation systems. Commun. ACM, 43:45–48.

Global Collaboration for Building new Learning Paradigms

Yoshiyori Urano

Graduate School of Global Information and Telecommunication Studies, Waseda University

Abstract

The recent rapid development of ICT has made it possible to realize various kinds of ICT- based Leaning paradigms. From the global point of views, however, there exists a great need of a new learning paradigm, namely, "u- Learning Paradigm".

We introduce recent activities on building new ICT-based learning paradigms in collaboration with people in Asian countries including the following cases;

(Case A: On-line Lecture and On-Demand Lectures)

With the sponsorship of three Funds (TAF, ICF and HBK) in Japan, Waseda University and PTIT/PTTC1 (Vietnam) have jointly carried out On-Line Lecture Project over the international ISDN, with PC-based AV teleconference systems since 1997.

As for On-Demand Lecture, AIC (Asia Info-communication Council)'s member countries (Vietnam, Malaysia, Thailand, Korea, China, Indonesia, Philippines) has developed learning materials for On-Demand Lectures on Mobile Communications and NGN (Next Generation Network).

(Case B: Learning Paradigm in Rural Areas)

We have jointly carried out HRD development programs, sponsored by APT (Asia Pacific Telecommunity), in Bario, Malaysia (2002), Ha Tinh, Vietnam (2003), Lombok, Indonesia (2004), Cebu, Philippines (2005), Phu Tho, Viet Nam (2006) and Phnom Penh, Cambodia (2007).

Based on the experience of these projects, we have proposed a new concept of e^n RAN (Rural Area Network)", where e^n refers not only to "electronic", associated with ICT, but also to "electronic", "energy", "environment", "ecology", "economy" and so on.

(Case C: Development of Chinese Language Learning System)

Due to the increasing popularity of the Chinese-Language and the lack of efficient methods in conventional education, the need for more effective computer-assisted learning system has been recognized. Since mobile phones services with flat-charge system have been getting popular in Japan, we have developed Chinese-Language Learning Systems on mobile phones which support learners to conduct self-learning, anytime and anywhere. In this system, a Virtual Assistant is newly proposed to provide learning-support functions, which give learners appropriate chances for their study, and further promote their learning motivation.

Through these collaborative projects on ICT-based Learning Paradigm, we have learned the following lessons;

(1) The style of ICT-based Learning should be different with each infrastructure in areas, regions and countries.

(2) ICT-based Learning would be evolving into u-Learning which is effective not only in Urban areas, but also in Rural areas.

(3) Universities should be one of important players in collaboration for building these learning paradigms. For example, universities could provide a forum where "Knowledge and Experience" in these projects are shared. Furthermore, foreign students can play a very important role in the sense that they can serve as the bridges between Japan and their respective country in carrying out the project smoothly and effectively.

We may conclude that it is very important for us to build new Learning Paradigms from "Global (Global & Local)" viewpoints, and in cooperation with people in the world.

Contents

- New ICT Trends in Japan
- From e-Japan to u-Japan,
- Case Studies
- Lessons learnt
- Conclusions

1. New ICT Trends in Japan

- 1995:   Basic Policy towards an Advanced Information and Tele-communications Society
- 2001:   e-Japan Policy; To become the world's most advanced IT nation by 2005
- 2004:   u-Japan Policy

2. u-Japan Policy

- U-Japan: The image of Next-Generation ICT Society in 2010
- 4U =   for You
- Ubiquitous, Universal, User-oriented, Unique
- u-Healthcare, u-Gaming, u-Learning

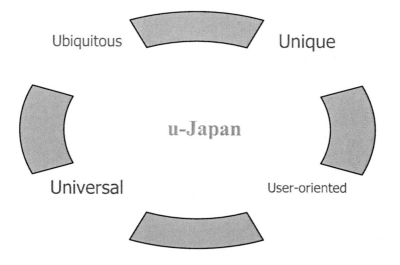

Ubiquitous · Unique

u-Japan

Universal · User-oriented

3. Our Approaches towards Building new Learning Paradigms

- 1996～Research on Distance Learning
- 1997～On-Line Lecture over ISDN in collaboration with PTIT, Hanoi
- 2001～AIC project of Web-based Learning Materials
- 2002～APT Project on HRD for ICT Development in Rural Areas
- 2003～On-Line Lecture over JICA Net
- 2004～Mobile Phone-based Chinese Language Learning System

4. Case A-1: On-Line Real-Time Lectures over International ISDN

- Sponsorship: three Japanese Funds: TAF, ICF and HBK
- Lectures over the international ISDN, with PC-based AV teleconference systems
- Contents; Introduction to Telecommunications, Broadband and Mobile Communications, Signal Processing, Policy, Regulation and Business

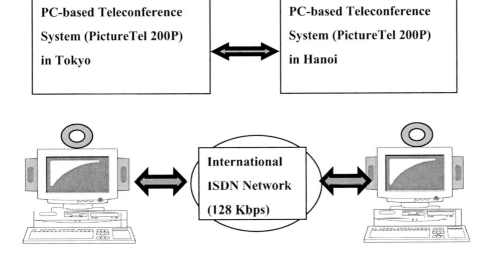

5. Case B: HRD for ICT Development in Rural Areas

- Sponsorship: APT (Asia Pacific Telecommunity)
- 2002: Bario, Malaysia
- 2003: Hatinh, Vietnam
- 2004: Lombok, Indonesia
- 2005: Cebu, Philippines

- 2006: Phutho, Viet Nam
- 2007: Phnom Penh, Cambodia

6. (Case B-1) e-Learning Paradigm in Bario, Malaysia

- Objectives: to promote the development of IC researchers and engineers through experiments of e-Learning over wireless LAN systems in rural areas.
- Constraints on power supply in Bario. A solution: small solar buttery.
- A new concept: "e^n-RAN (Rural Area Network)", where e^n refers not only to "electronic", associated with ICT, but also to "electric", "environment" and "ecology" and so on.

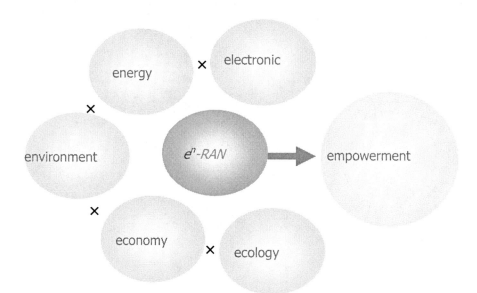

7. (Case B-2) e-Learning Paradigm in Phu Tho, Vietnam

- Objectives: to reduce the digital divide in rural areas through training of human resources for ICT.

- A pilot system with WiMAX for Deployment of Broadband Services and e-Applications (e-Learning and Information Sharing by GPS) in Phutho.

- For sustainability, win-win relationship among members of projects

8. (Case C) Development of Mobile Phone based Chinese Learning System

- Backgrounds:
    - (1) The Demands on Learning Chinese Language due to Advance-
      ment of Tradesbetween China and Japan
    - (2) The Current Status of Mobile Phone Services in Japan

- Mobile Phone based Chinese Learning System

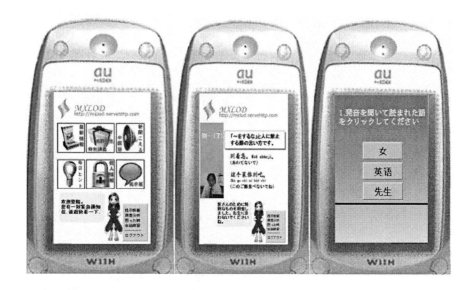

9. Lessons learned (1)

- The style of e-Learning should be different with each infrastructure in areas, regions and countries.
- The style of ICT-based Learning should be different with each infra-structure in areas, regions and countries.
- ICT-based Learning would be evolving into u-Learning which is effective not only in Urban areas, but also in Rural areas.

10. Lessons learned (2)

- The Universities should play important roles in collaboration for building these learning paradigms.
- Foreign students can be important players in the sense that they can serve as the bridges between Japan and their respective country in carrying out the project smoothly and effectively.

12. Conclusion

- It is important to build new Learning Paradigms from "Glocal (Global & Local) viewpoints and in cooperation with people in the world.

Common and User-Friendly Text Input Interfaces for Asian Syllabic Languages on Mobile Devices

Ye Kyaw Thu

Graduate School of Global Information and Telecommunication Studies

Waseda University

Tokyo, JAPAN

Abstract I believe that text typing on small mobile devices will become more popular and necessary communication in Asian developing countries such as Myanmar (Burma), Bangladesh, Nepal, Bhutan, Laos and Cambodia etc. In these countries, however, there is no efficient and user-friendly text input method for mobile devices yet. Asian languages are syllabic languages that derived from Indic script or Brahmi around BC third century. And thus, Myanmar language or Burmese, Bengali, Nepali, Dzongkha (language of Bhutan), Lao and Khmer have common writing natures with Indian languages such as Hindi, Marathi, Punjabi and Tamil etc. But current mobile devices key-mapping or text input methods such as multi-tap or T9 are based on English and not directly applicable to syllabic languages, because writing natures of Asian syllabic languages are different and have larger numbers of characters than English alphabets (e.g. Khmer has triple numbers of characters (i.e. 74) of English). My research looks for common and user-friendly keyboard mappings and text input methods for Asian syllabic languages based on their word formation or writing natures for mobile devices. I have already proposed 1) "Positional Mapping (PM)" [1], [2], [3], [4], [5], [6] for mobile phone or Personal Digital Assistant (PDA) keyboard mapping 2) "Positional Gesture (PG)" [7], [8] for gesture text input interface and 3) "Positional Prediction (PP)" [9], [10] for consonant cluster predictive text input.

Contents

1. Introduction

Here, I briefly explain writing systems of Asian syllabic languages (e.g. Myanmar, Khmer and Bengli) and their common nature.

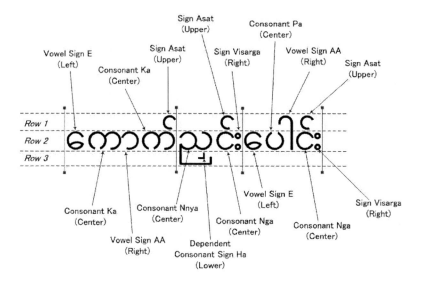

Fig.1 Myanmar word "Kauk Nyin Baung" (festive red rice)

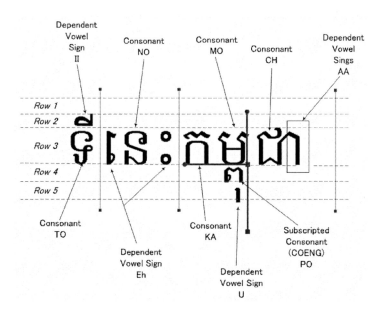

Fig.2 Khmer sentence "Ti ni Kampuchar" (This is Cambodia)

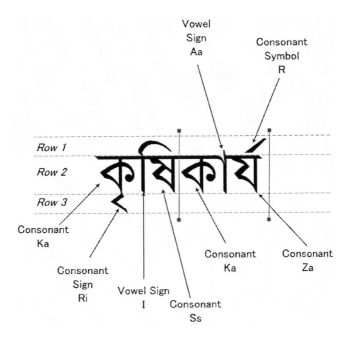

Fig.3 Bangla word "Krissikaarza" (Agriculture)

> ➢ Through my study, I have noticed that there are many common characteristics among the writing system of Asian syllabic language.
> ➢ Most of them are related to each other, and their writing system largely depends on adding left, right, upper and lower characters to a consonant (i.e. consonant clusters or syllable).
> ➢ Here, left, right, upper and lower characters mean dependent vowels, directives and subscript consonants that are always written with a consonant.
> ➢ I present the similar nature of "Logical Structure or Combination Structure" of Myanmar, Bangla and Khmer languages' consonant clusters. (see Fig.4, Fig.5 and Fig.6).

2. Logical Structure of Asian Syllabic Languages

Upper
+
Left + Left + Consonant + Right + Right + Right
+
Lower

Fig.4 Myanmar consonant clusters combination structure

Upper
+
Left + Consonant + Right
+
Lower

Fig.5 Bangla consonant clusters combination structure

Upper
+
Upper
+
Left + Consonant + Right + Right
+
Lower
+
Lower

Fig.6 Khmer consonant clusters combination structure

I investigate keyboard mapping, gesture text input and consonant cluster or syllable prediction method for Asian syllabic languages based on their word formation or structure of a consonant cluster.

3. Positional Mapping (PM)

Positional Mapping (PM) is a concept of keyboard or keypad mapping for mobile phone based on common characteristics of writing systems of Asian syllabic languages (see Fig.7) [1], [2], [3], [4], [5], [6].

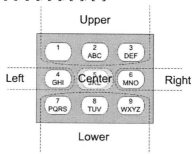

Fig.7 Concept of Positional Mapping

[For Myanmar Characters]

1 = ပါ,တယ်,လား,လဲ,သည်,၏,တို,ဖြင့်,ဟို,ဒီ,အဲ,ငါ,နှင့်,သူ,မင်း,ညည်း,ရှင့်,ရှင့်,သင်,ကျုပ်,ကျွန်တော်,ကျွန်မ,ကျနော်,ကျမ,ကျွန်ုပ်,
ကျုပ်,ပြား,ခွဲ,နာရီ,မိနစ်,ရာ,ထောင်,သောင်း,သိန်း,သန်း,ကုဋေ, etc.

2 = ◌ဲ ◌ဲ ◌ဳ ◌ဴ ◌ဲ ◌ဲ

3 = ◌ူ ◌ူ ◌ူ ◌ူ ◌ူ

4 = ေ◌,ြ◌

5 = က,ခ,ဂ,ဃ,င,စ,ဆ,ဇ,ဈ,ည,ဍ,ဌ,ဎ,ဏ,တ,ထ,ဒ,ဓ,န,ပ,ဖ,ဗ,ဘ,မ,ယ,ရ,လ,ဝ,သ,ဟ,ဠ,အ

6 = ◌ာ,◌ျ,◌ႏ

7 = ◌ိ,◌ီ,◌ႈ,◌ႎ

8 = ◌ံ,◌ိံ,◌ိ

9 = space,◌ႂ,◌ဳ

0 = ◌ၚ,◌ဲ,◌ဲ,◌ဲ,◌ဲ,◌ဲ,◌ၬ,◌ၚ,◌ၚ,◌ဲ,◌ဲ,◌ဲ,◌ဲ,◌ဲ,◌ၚ,◌ၚ,◌ၚ,◌ဲ,◌ဲ,◌ၚ

* = ကွ,ညှ,ၐ,ၑ,ဩ,ၑ,၍,ၒ,ရင်း,ၐ,ဪ,ၐ,ၐ,ၑ,ၑ,ည,ၐ

= @,.,,,:,;,_,-,$,%,!,?,&,#,+,*,=,/,pipe,~,",',^,`,(,),<,>,[,],{,}

Fig.8 Example key assignment for Positional Mapping

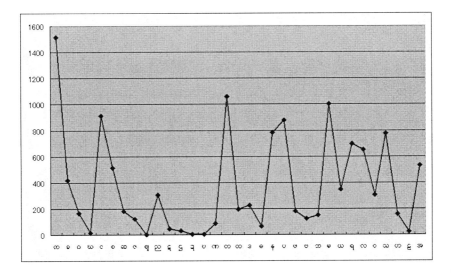

Fig.9 Usage frequency of Myanmar consonants Ka to A

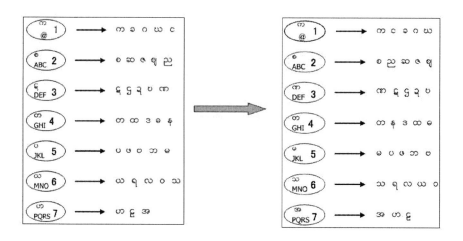

Fig.10 Ordered Myanmar consonants by usage frequency

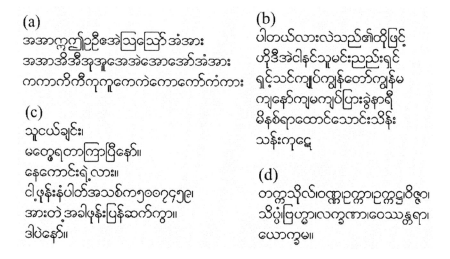

Fig.11 Myanmar text for user experiments
(a) vowel combinations (b) frequently used Myanmar words (c) SMS email (d) Pali words

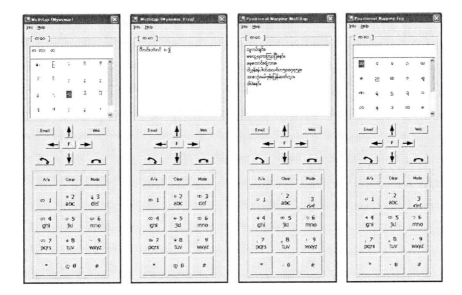

Fig.12 Simulation program for Positional Mapping on mobile phone

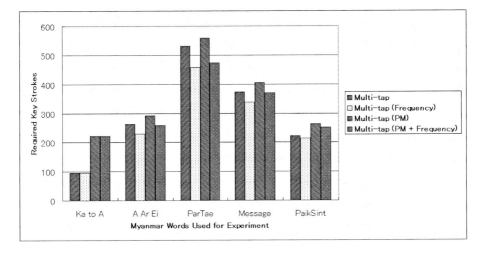

Fig.13 Keystrokes per Characters (KSPC) comparison for 4 models of PM

Input Method	User1	User2	User3	User4	User5
Multi-tap	6:52	9:33	6:15	5:15	4:26
Multi-tap (Frequency)	5:42	6:56	5:58	4:47	5:13
Multi-tap (PM)	4:29	4:43	4:16	4:24	4:20
Multi-tap (PM Frequency)	4:17	4:00	4:18	4:03	4:04

The fastest typing speed for "Multi-tap" = 4 min 26 sec

The fastest typing speed for "Multi-tap + Frequency" = 4 min 47 sec

The fastest typing speed for "Positional Mapping" = 4 min 16 sec

The fastest typing speed for "Positional Mapping + Frequency" = 4 min 00 sec

The slowest typing speed for "Multi-tap" = 9 min 33 sec

The slowest typing speed for "Multi-tap + Frequency" = 6 min 56 sec

The slowest typing speed for "Positional Mapping" = 4 min 43 sec

The slowest typing speed for "Positional Mapping + Frequency" = 4 min 18 sec

Fig.14 Typing speed comparison for 4 models of PM

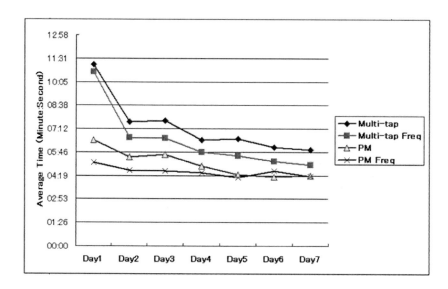

Fig.15 Typing speed improvements of slowest user for 4models

Fig.16 Positional Mapping prototype for numeric keypad and PDA

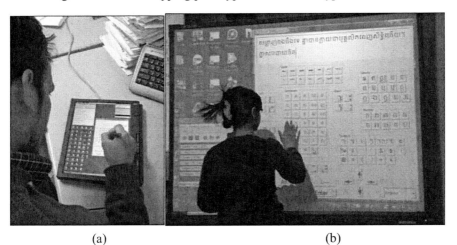

(a)   (b)

Fig.17 Photos from user study
(a) user study for Bangla with tablet PC
(b) user study for Khmer with electronic whiteboard

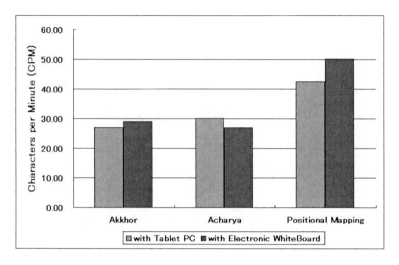

Fig.18 Character per Minute (CPM) comparison
for Akkhor, Acharya and Positional Mapping software keyboards

Discussion on Positional Mapping

Advantages

➢ User-friendly key mapping (simple and easy to memorize for users)
➢ Positional Mapping gives the highest typing speed (25 % faster than our proposed Multi-tap model)
➢ Positional Mapping idea can be applied not only to Myanmar languages but also to other similar phonetic based languges such as Khmer, Thai, Hindi and Bangla etc.
➢ All of the participants choose "Multi-tap (PM)" or "Multi-tap (PM + Frequency)" as the best input interface
➢ Positional Mapping is applicable not only for mobile phones but also for pen based mobile devices such as PDA, Tablet-PC etc.

Disadvantage

➢ Positional Mapping requires more keystrokes (the reason is that all of the 33 consonants are assigned as a list only on No.5 Key)

4. Positional Gesture (PG)

In the Next Generation Network (NGN), various kinds of advanced smart mobile terminals will be used for various communication services. I believe that text typing on small mobile devices will become more popular than it is today and also necessary for developing countries such as Myanmar (Burma), Cambodia and Bangladesh etc.

In these countries, however, there is no proper or easy text input method for mobile devices yet.

Positional Gesture Text Input is a novel concept of text input for syllabic scripts like Myanmar, Khmer and Bangla languages. Text input of syllabic scripts poses a unique challenge because many syllabic characters are formed by combinations of consonants, dependent vowel signs, tones and subscript consonants etc. And thus, text input for syllabic scripts is still difficult even with PC keyboards.

I proposed very simple gesture recognition for syllabic scripts text input based on their writing natures. It is accessible even for first time users and applicable for many mobile computing devices such as tablet PCs, mobile phones, PDAs and portable game players etc.

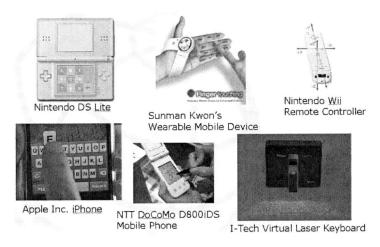

Fig.19 Human communication with various kinds of mobile devices

Related Works

Unistroke
[Goldberg 93]

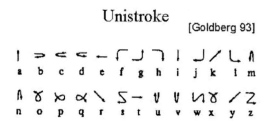

Graffiti

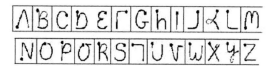

Proposed by Palm Inc.

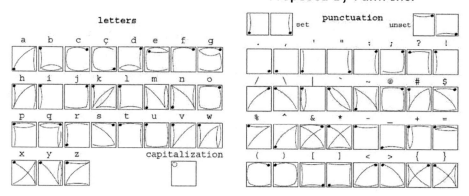

Proposed by Wobbrock, J.O. (2006)

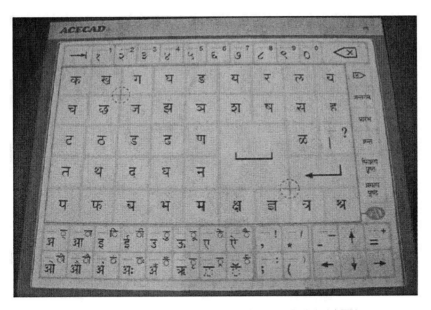

Proposed by HP Lab. India at HCII 2005

Concept of Positional Gesture

Positional Gesture (PG) is a simple gesture text input method for mobile devices based on common characteristics of syllabic scripts writing system [7], [8].

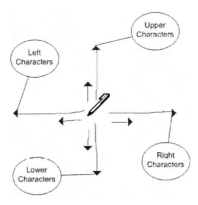

Fig.20 Concept of Positional Gesture text input

Gesture Commands	Character Assignments
Left (long)	Left characters ("ၘ-", "ြ-")
Right (long)	Numbers ("၁", "၂", "၃", "၄", "၅" etc.)
Up (long)	Symbols ("@", "!", "&", "$", "%" etc.)
Down (long)	Subscript consonants ("ၐ", "ၟ", "ၔ", "ၒ", "ၕ" etc.)
Left (short)	Consonant characters ("က", "ခ", "ဂ", "ဃ", "င" etc.)
Right (short)	Right characters ("ျ", "ာ", "ီ", "ြ", "။")
Up (short)	Upper characters ("ိ", "ီ", "ဲ" etc.)
Down (short)	Lower characters ("ု", "ူ", "ွ", "ှ", "ဲ" etc.)

Gesture Commands	Character Assignments
Left (long)	Left, Right characters (" េ្", "េ្", "េា" etc.)
Right (long)	Numbers ("១", "២", "៣", "៨" etc.)
Up (long)	Symbols ("!", "", "?", "!", "#" etc.)
Down (long)	Independent vowels, Symbols and frequently used characters ("ឥ", "ឭ", "ៗល។", "ៈ" etc.)
Left (short)	Left characters ("េ", "ៃ", "ៃ", "េ")
Right (short)	Right characters ("ា", "ះ", "ឺ", "ៈ")
Up (short)	Upper characters ("ី", "ឹ", "ឺ", "ឺ" etc.)
Down (short)	Lower characters ("ុ", "ូ", "ួ" etc.)
Double Click	Consonant characters ("ក", "ខ", "គ", "ឃ", "ង" etc.)

Table.1 Gesture commands for (left) Myanmar and (right) Khmer

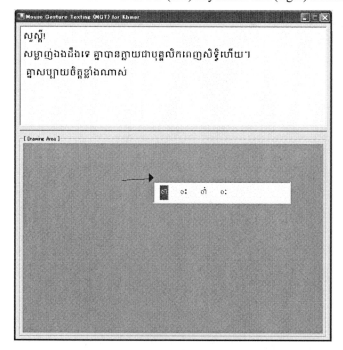

Fig.21 Positional gesture prototype for Khmer

Experiment Procedures

➢ Explaining the concept of Positional Gesture text input
➢ Making demonstration of text input with Positional Gesture prototype and NiDA software keyboard
➢ Allowing 10 minutes practice time for each user to learn text input with Positional Gesture prototype and software keyboard
➢ Recording users' typing speeds of short Khmer message for 5 trial times (including error correction time)
➢ Getting users feedback for our Positional Gesture prototype and NiDA software keyboard with small questionnaires

Fig.22 User study with native participants for Myanmar and Khmer PG prototypes

NiDA Software Keyboard for Comparison

NiDA Software Keyboard Layout (Unshifted Mode)

NiDA Software Keyboard Layout (Shift Mode)

NiDA Software Keyboard Layout (Right Alt Mode)

User Study Result for Myanmar Language

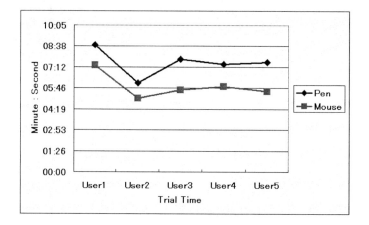

Fig.23 Typing speed improvements of 5 Myanmar users
for Positional Gesture with mouse and pen

User Study Result for Khmer Language

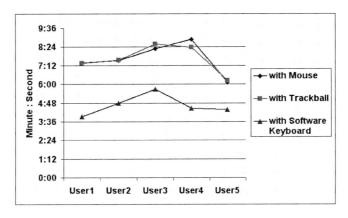

Fig.24 Typing speed comparison of 5 Cambodian users
with mouse, trackball and software keyboard

Characters per Minute (CPM)

For Myanmar Language
➢ Positional Gesture with mouse: 19.27 CPM
➢ Positional Gesture with pen: 14.29 CPM
➢ Win Myanmar Software Keyboard: 42.8 CPM

For Khmer Language
➢ Positional Gesture with mouse: 17.69 CPM
➢ Positional Gesture with trackball: 17.72 CPM
➢ Khmer Software Keyboard: 29.13 CPM

Result of Likert Scales by Myanmar Users

Likert Scales (range 1-5)	PG with Pen	PG with Mouse
Difficult-Easy	3.4 (0.89)	4.6 (0.55)
Painful-Enjoyable	3.6 (0.55)	3.4 (1.14)
Slow-Fast	3.0 (0)	4.2 (0.84)
Dislike-Like	4.4 (0.89)	4.6 (0.55)

Mean (Standard Deviation) Responses by 5 Myanmar Users for 5-point Likert Scale Questions

Result of Likert Scales by Cambodian Users

Likert Scales (range 1-5)	PG with Trackball	PG with Mouse	Software Keyboard
Difficult-Easy	2.0 (1.22)	3.2 (0.84)	4.2 (1.30)
Painful-Enjoyable	2.6 (1.14)	3.6 (0.89)	3.8 (1.10)
Slow-Fast	2.0 (0.71)	3.4 (1.14)	3.8 (1.10)
Dislike-Like	2.8 (1.79)	4.0 (0.71)	4.0 (1.22)

Mean (Standard Deviation) Responses by 5 Cambodian Users for 5-point Likert Scale Questions

Discussion on Positional Gesture

➢ The proposed gesture idea is a very simple, user-friendly and possible text input method on computing devices.
➢ Moreover, the concept is applicable to many pointing devices or input devices such as mouse, TouchPad, TrackPoint, trackball, pen with tablet, touch screen and data glove etc.
➢ Positional Gesture with touch screen interface can be the best user interface.
➢ Positional Gesture text input concept can be applicable for disabled or handicapped person
➢ As a next step, we plan to extend it to Thai and Hindi, and make an analysis of typing error rate as well.

5. Positional Prediction (PP)

➢ Many alternative text entry methods exist for English, Japanese and Chinese etc., however, there is no efficient method for Myanmar language mobile devices.
➢ There is no predictive text entry method for Myanmar language, and no mobile phone company supports Myanmar language text input method yet.
➢ Therefore, I propose user-friendly predictive text entry method for Myanmar language.

Concept of Positional Prediction Text Input

Positional Prediction is a new consonant cluster (i.e. one consonant + other characters especially vowels) prediction concept based on positional vowel information [9], [10].

Example: Steps to type a word "human" in Myanmar language

| (Consonant "La") | (Positional Information of Vowel) | (Selection from the Possible Combination List of "La") | (Consonant Cluster "Lu") |

Process Flow of Positional Prediction

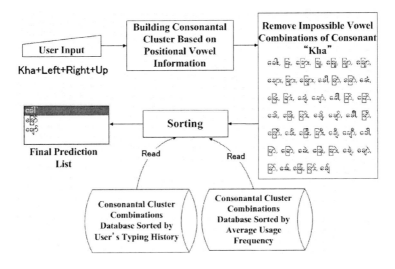

Character Combination Error Checking

$$\sum_{i=1}^{m} {}_m C_i = \sum_{i=1}^{m} \frac{m(m-1) \times (m-2) \cdots (m-i+1)}{i(i-1)(i-2) \cdots 1}$$

$$mf + ml = \sum_{i=1}^{m} {}_m C_i$$

m = total number of Myanmar characters in a word
mf = number of meaningful combination patterns
ml = number of meaningless combination patterns

Positional Prediction Text Entry Prototype for Myanmar

Photos of Initial User Study in Myanmar and Japan

Even 5 years old Myanmar children can type their names with Positional Prediction text input prototype

Result of Initial User Study (with Mouse) for Myanmar

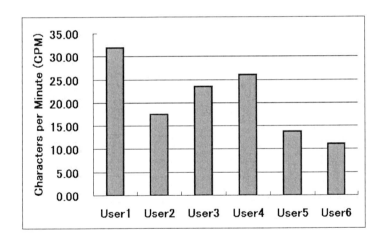

Result of Initial User Study (with Pen) for Myanmar

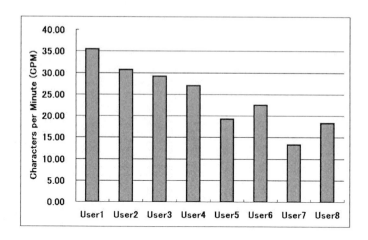

Result of Initial User Study (with different input devices) for Myanmar

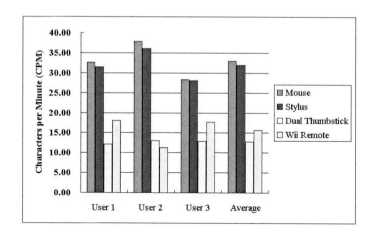

~41.57 CPM - with mouse
~41.84 CPM - with stylus pen
~14.76 CPM - with dual thumbstick game pad
~20.52 CPM - with Nintendo Wii remote controller

Applying Positional Prediction Text Input Concept for Khmer

For example,

[Ka (ក) + ←] for "រក", "រក", "រក" and "រក"

[Ka (ក) + →] for "កា", "កាំ", "កះ", "កោ", "កៅ", "កៀ", "កៀ" and "ក:"

[Ka (ក) + ↑] for "កិ", "កី", "កឹ", "កឺ", "កុ", "កូ", "កួ", "កើ", "កឿ" and "កៃ"

[Ka (ក) + ↓] for "ក្", "ក្", "ក្", "ក្" and "ក:"

[Ka (ក) + ↓ + ↑] for "ក្", "ក្" etc.

Results of Initial User Study for Khmer

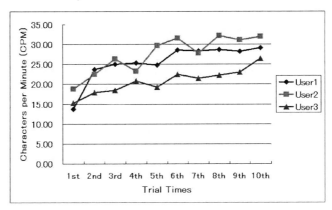

By mouse:
Average: 5 min 44 sec (24.62 CPM)
Fastest: 4 min 11 sec (32.27 CPM)
Slowest: 9 min 48 sec (13.78 CPM)

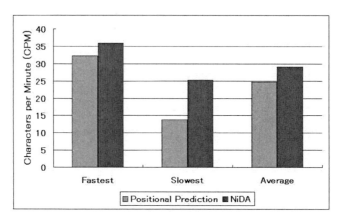

Fig.25 Fastest CPM, slowest CPM and average CPM
between Positional Prediction and NiDA software keyboard

Likert Scales (range 1-5)	U1	U2	U3	Average
Difficult-Easy	4	4	4	4
Painful-Enjoyable	4	5	3	4
Slow-Fast	5	4	2	3.67
Dislike-Like	5	4	3	4

Users' evaluations for Positional Prediction text input
(U1, U2 and U3 mean User1, User2 and User3 respectively, and higher values are better.)

Prototype for PDA and Small Mobile Devices

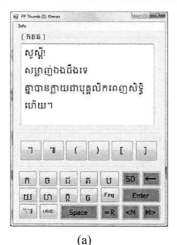

(a)   (b)

Fig.26 (a) PP_Thumb Prototype for PDA (b) Prototype for iPOD and Game Pad

Discussion on Positional Prediction

➢ The merit of the proposed predictive text input method is that even first time users can type Myanmar and Khmer sentences with appropriate typing speed.
➢ My approach is applicable for various kinds of mobile devices.

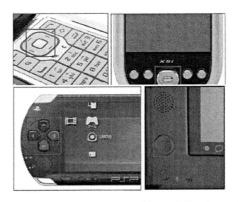

Fig.27 Four directional arrow keys on Nokia mobile phone, Dell PDA,
Sony PSP game player and XO laptop

➢ Positional Prediction is extendable to other similar syllabic languages such as Nepali (language of Nepal), Thai (language of Thailand) and Hindi (one of the languages of India) etc.
➢ I am developing prototypes for Bangla and Thai languages. I also plan to apply this concept for other syllabic languages.
➢ I am planning to make user experiments with real mobile phones in the near future.

6. Conclusion and Future Work

➢ Keyboard Mapping and creating text input interfaces of Asian languages based on consonant cluster construction or word formation is a user-friendly Human-Computer Interface.
➢ I have already created some prototypes and held initial user studies for Myanmar, Khmer and Bangla languages, and am now planning to make analysis with other similar languages.
➢ I plan to make Positional Gesture text input interfaces user study with handicapped people in the near future.

7. Acknowledgements

I would like to thank my supervisor Prof. Yoshiyori URANO for his great support for my research work as well as scholarship offered by Japanese Ministry of Education, Culture, Sports, Science and Technology throughout my doctoral course.

8. Publication List for PM, PG and PP

1. Ye Kyaw Thu and Yoshiyori URANO, "Positional Mapping Multi-tap for Myanmar Language", in *proceeding of the 12th International Conference on Human–Computer Interaction (HCI International 2007)*, July 22~27, 2007, Beijing International Convention Center, Beijing, P.R.China, (Proceedings Volume 2, LNCS_4551, ISBN: 978-3-540-73106-1) Page 486 – 495

2. Ye Kyaw Thu and Yoshiyori URANO, "Positional Mapping Myanmar Text Input Scheme for Mobile Devices", in *proceeding of the International Conference on Human Computer Interaction with Mobile Devices and Services (MobileHCI 07)*, September 9~12, 2007, Singapore Polytechnic Auditorium, Singapore, Page 153 – 160

3. Ye Kyaw Thu and Yoshiyori URANO, "Positional Mapping: Keyboard Mapping Based on Characters Writing Positions for Mobile Devices", in *proceeding of the 9th International Conference on Multimodal Interfaces (ICMI 2007)*, November 12~15, 2007, Nagoya, Japan, Page 110 – 117

4. Ye Kyaw Thu, MD. Monzur Morshed and Yoshiyori URANO, "Positional Mapping for Bangla Mobile Phone", *Forum on Information Technology (FIT 2007)*, September 5~7, 2007, Chukyo University, Toyota Campus, Toyota, Japan, Page 427 – 430

5. Ye Kyaw Thu and Yoshiyori URANO, "Myanmar Text Typing with Numeric Keypad (Applying the concept of Positional Mapping)", *Proceedings of the 70th IPSJ National Conference*, March 13~15, 2008, Tsukuba University, Tsukuba, Japan, Page 4-1 – 4-2

6. Ye Kyaw Thu, Ouk Phavy and Yoshiyori URANO, "Positional Mapping Software Keyboard for Syllabic Scripts (Study with Bangla and Khmer Languages)", in *proceeding of the 6th International Conference on Computer Applications (ICCA2008)*, February 14~15, 2008, Yangon, Myanmar, Page 150-157

7. Ye Kyaw Thu, Ouk Phavy and Yoshiyori URANO, "Positional Gesture: Simple Gesture Text Input for Khmer", *the 127th Human Computer Interaction Symposium 2008*, January 31 to February 1, 2008, Faculty Club (Gakushi Kaikan), Hiroshima University, Higashi Hiroshima, Japan, SIGHCI, Information Processing Society of Japan (IPSJ) SIG Technical Reports, 2008 – HCI – 127, Page 115–122

8. Ye Kyaw Thu, Ouk Phavy and Yoshiyori URANO, "Positional Gesture for Advanced Smart Terminals: Simple Gesture Text Input for Syllabic Scripts Like Myanmar, Khmer and Bangla", *in proceedings of the first ITU-T Kaleidoscope Academic Conference*, May 12~13, 2008, Geneva, Switzerland, Page 77 – 84

9. Ye Kyaw Thu, Ouk Phavy, Yoshiyori URANO and Mitsuji MATSUMOTO, "Positional Prediction for Khmer Language (Cluster Predictive Text Entry Method in Mobile Devices)", IIEEJ *4th Mobile Image Research Meeting*, March 4, 2008, International Conference Center, Waseda University, Tokyo, Japan, Page 5 – 10

10. Ye Kyaw Thu and Yoshiyori URANO, "Positional Prediction: Consonant Cluster Prediction Text Entry Method for Burmese (Myanmar Language)", *in proceeding of the 26th ACM Conference on Human Factors in Computing systems (CHI 2008)*, April 5~10, 2008, Florence, Italy, Page 3783-3788

WASEDA UNIVERSITY

Research and Development of Next-Generation Free-Space Optical Communication Systems

Kamugisha Kazaura, Chen Yanru, Christian Sousa and Mitsuji Matsumoto

Matsumoto Lab, GITS/GITI, Waseda University

University of Chile

WASEDA UNIVERSITY

WASEDA UNIVERSITY

Abstract

Significant progress has been made in recent years in the research, development and deployment of affordable, ubiquitous and always-on broadband wireless access technologies that meet the demands and requirements of business, academic and residential end user markets. One of the emerging technologies for broadband wireless connectivity which has also been receiving growing attention is free-space optical (FSO) communication. This talk will describe two NiCT sponsored projects being carried out at Matsumoto Lab (MMLab) in GITS, Waseda Univ. on research and development of Next Generation FSO systems as well as development of new advanced dense wavelength division multiplexing (DWDM) Radio-on-FSO (RoFSO) communication systems. The design concept and experimental evaluation of the systems are outlined.

Also covered in this session are two other research activities also carried out in MMLab. First is research on developing an epidemic P2P computing system which is motivated by requirement of processing of high-resolution videos. The next is investigation on possibility of adding support for multi-touch input devices with the aim of seamlessly using common functions in traditional graphical user interfaces.

WASEDA UNIVERSITY

2/69

WASEDA UNIVERSITY

Contents

- NiCT project research
  - Research and development of Next Generation FSO communication systems
  - Research and development of New Advanced DWDM RoFSO communication systems

- Summary

- Introduction to GITS and MMLab

WASEDA UNIVERSITY

3/69

WASEDA UNIVERSITY

NiCT Sponsored Research Projects

NiCT: National Institute of Information and Communications Technology

WASEDA UNIVERSITY

4/69

WASEDA UNIVERSITY

Project I:
Research and Development of Next Generation FSO communication systems

WASEDA UNIVERSITY — 5/69 — Matsumoto Mitsuji Laboratory, OPTICAL & WIRELESS LAB

WASEDA UNIVERSITY

Contents

- Introduction
  - Project details
  - Wireless systems, technologies and standards
- Overview of FSO systems
  - What is FSO, application, technology etc
  - Factors influencing performance of FSO systems
- Next Generation FSO system
  - Transceiver design and performance
- Performance evaluation of Next Generation FSO system
- Summary

WASEDA UNIVERSITY — 6/69 — Matsumoto Mitsuji Laboratory, OPTICAL & WIRELESS LAB

WASEDA UNIVERSITY

Introduction
Project details

- Period:
  - April 2004 ~ March 2006 (2 years)
- Collaborating entities
  - Waseda Univ.
  - Hamamatsu Photonics
  - NiCT
  - Olympus Corporation
  - Showa electric
  - Koito Industries
  - Canon
  - Sponsor NiCT

WASEDA UNIVERSITY — 7/69 — Matsumoto Mitsuji Laboratory, OPTICAL & WIRELESS LAB

WASEDA UNIVERSITY

Introduction cont.
Wireless communication systems

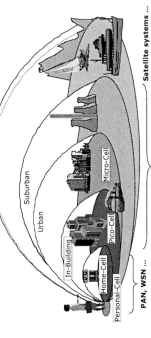

Global

Suburban

Urban

Micro-Cell

In-Building

Pico-Cell

Home-Cell

Personal-Cell

PAN, WSN ...

FSO, Cellular systems, WiMAX ...

Satellite systems ...

WASEDA UNIVERSITY — 8/69 — Matsumoto Mitsuji Laboratory, OPTICAL & WIRELESS LAB

WASEDA UNIVERSITY

Introduction cont.
Wireless communication technologies and standards

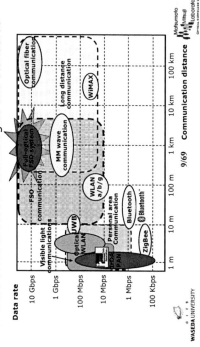

9/69   Communication distance

- Optical fiber communication
- Long distance communication
- WIMAX
- Full-optical FSO system
- MM wave communication
- FSO communication
- Visible light communications
- WLAN a/b/g
- Optical UWB WLAN
- Personal area Communication
- Bluetooth
- CAN
- ZigBee

Data rate: 10 Gbps, 1 Gbps, 100 Mbps, 10 Mbps, 1 Mbps, 100 Kbps

Communication distance: 1 m, 10 m, 100 m, 1 km, 10 km, 100 km

---

WASEDA UNIVERSITY

Introduction cont.
FSO roadmap

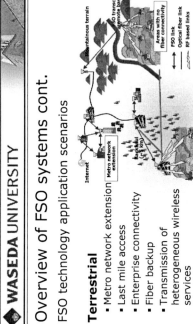

Wireless BB environment

Cooperation of fiber comm.

FSO10G (WDM)

100G-Ether

FTTH 1G

Radio on FSO

FSO 2.5G (Eye safe)

FWA 50M

FWA 46M (3G)

10G-Ether standard

FSO 1G

Indoor FSO (P-MP)

11a

24M

WDM

SONET (trunk line)

1G-Ethernet standard

FSO 100M

FTTH

FWA 10M (22 & 26G)

IEEE802.11b

ADSL

CATV, cellular phone   Analog FSO system

FSO 10M

Indoor FSO (P-MP)

FWA 1.5M (22 & 26G)

ISDN

Video use

Data rate: 1T, 100G, 10G, 1G, 100M, 10M, 1M, 100K

~1995, ~2000, 2001, 2002, 10/69 2003, 2004, 2005

WASEDA UNIVERSITY

Overview of FSO systems

FSO is the transmission of modulated visible or infrared (IR) beams through the atmosphere to obtain broadband communication.

RoFSO contains optical carriers modulated in an analogue manner by RF sub-carriers.

Merits

- Secure wireless system not easy to intercept
- Easy to deploy, avoid huge costs involved in laying cables
- License free
- Possible for communication up to several kms
- Can transmit high data rate

De merits

- High dependence on weather condition (rain, snow, fog, dust particles etc)
- Can not propagate through obstacles
- **Susceptible to atmospheric effects (atmospheric fluctuations)**

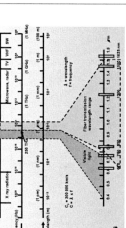

11/69   Electromagnetic spectrum

WASEDA UNIVERSITY

Overview of FSO systems cont.
FSO technology application scenarios

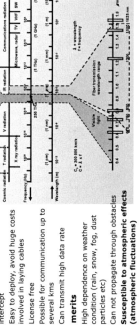

Terrestrial

- Metro network extension
- Last mile access
- Enterprise connectivity
- Fiber backup
- Transmission of heterogeneous wireless services

12/69

WASEDA UNIVERSITY

Overview of FSO systems cont.

FSO technology application scenarios

Space

- Inter-satellite communication (cross link
- Satellite to ground data transmission (down link)
- Deep space communicatio

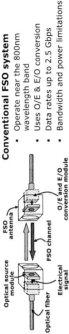

13/69

WASEDA UNIVERSITY

Overview of FSO systems cont.

Conventional FSO system

- Operate near the 800nm wavelength band
- Uses O/E & E/O conversion
- Data rates up to 2.5 Gbps
- Bandwidth and power limitations

Next generation FSO system

- Uses 1550nm wavelength
- Seamless connection of space and optical fiber.
- Multi gigabit per second data rates (using optical fiber technology)
- Compatibility with existing fiber infrastructure
- Protocol and data rate independent

(a) Conventional FSO system

(b) New full-optical FSO system

14/69

WASEDA UNIVERSITY

Overview of FSO systems cont.

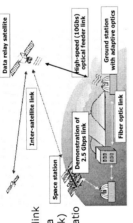

(c) Advanced DWDM RoFSO system

Advanced DWDM RoFSO system

- Uses 1550nm wavelength
- Transport multiple RF signals using DWDM FSO channels
- Realize heterogeneous wireless services e.g. WLAN, Cellular, terrestrial digital TV broadcasting etc

15/69

WASEDA UNIVERSITY

Overview of FSO systems cont.

Challenges in design of FSO systems

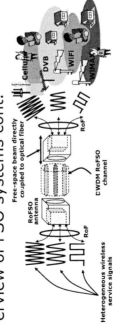

Wide beam FSO systems

- Beam divergence in terms of several milliradians
- Easy to align and maintain tracking
- Less power at the receiver (the wider the beam the less power)

Narrow beam FSO systems

- Beam divergence in terms of several tens of microradians
- Difficult to align and maintain tracking
- More optical power delivered at the receiver

The narrow transmission of FSO beam of makes alignment of FSO communication terminals difficult than wider RF systems.

16/69

Slide 17/69

WASEDA UNIVERSITY

Overview of FSO systems cont.
FSO system performance related parameters

FSO Performance

Internal parameters (design of FSO system)
- Optical power
- Wavelength
- Transmission bandwidth
- Divergence angle
- Optical losses
- BER
- Receive lens diameter & FOV

External parameters (non-system specific parameters)
- Visibility
- Atmospheric attenuation
- Scintillation
- Deployment distance
- Pointing loss

WASEDA UNIVERSITY

Matsumoto Mitsuji Laboratory
OPTICAL & WIRELESS LAB

17/69

Slide 18/69

WASEDA UNIVERSITY

Overview of FSO systems cont.
FSO system performance related parameters

- Key issue in FSO systems deployment is link availability.

- Link availability comprises of:
  - Equipment reliability ⎤
  - Network design   ⎬ → Quantifiable
  - Atmospheric effects ⎦ → Statistical data obtained by measurements or theoretical calculations

WASEDA UNIVERSITY

Matsumoto Mitsuji Laboratory
OPTICAL & WIRELESS LAB

18/69

Slide 19/69

WASEDA UNIVERSITY

Overview of FSO systems cont.
Factors influencing performance of FSO systems
Visibility under different weather conditions

(a)

(b)

(c)

Clear day
Visibility > 20km
Attenuation: 0.06 ~ 0.19 db/km

Cloudy day
Visibility: ~ 5.36 km
Attenuation: 2.58 db/km

Rain event
Visibility: ~ 1.09 km
Attenuation: 12.65 db/km

Visibility greatly influences the performance of FSO systems e.g. fog, rain, snow etc significantly decrease visibility

WASEDA UNIVERSITY

Matsumoto Mitsuji Laboratory
OPTICAL & WIRELESS LAB

19/69

Slide 20/69

WASEDA UNIVERSITY

Overview of FSO systems cont.
Factors influencing performance of FSO systems
Atmospheric effects

Atmospheric turbulence has a significant impact on the quality of the free-space optical beam propagating through the atmosphere.

Transmit power

Received power

Time

Beam wander

Other effects include:
- beam broadening and
- angle-of-arrival fluctuations

Time

Scintillation

Time

Combined effect

Time

Reduces the optical beam power at the receiver point and causes burst errors

WASEDA UNIVERSITY

Matsumoto Mitsuji Laboratory
OPTICAL & WIRELESS LAB

20/69

-223-

Slide 21 (top-left)

Overview of FSO systems cont.

Atmospheric effects suppression techniques

- Aperture averaging
  - Reducing scintillation effects by increasing the telescope collecting area.
- Adaptive optics
  - Measure wavefront errors continuously and correct them automatically.
- Diversity techniques
  - Spatial diversity (multiple transmitters and/or receivers)
  - Temporal diversity (signal transmitted twice separated by a time delay)
  - Wavelength diversity (transmitting data at least two distinct wavelengths)
- Coding techniques
  - Coding schemes used in RF and wired communications systems.

Matsumoto
Mitsuji Laboratory
OPTICAL WIRELESS LAB

21/69

WASEDA UNIVERSITY

Slide (top-right)

Overview of FSO systems cont.

Atmospheric effects suppression technique using adaptive optics (AO)

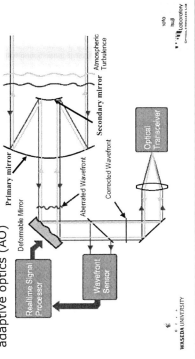

Primary mirror
Deformable Mirror
Secondary mirror
Atmospheric Turbulence
Aberrated Wavefront
Corrected Wavefront
Optical Transceiver
Realtime Signal Processor
Wavefront Sensor

Toto
Itsuji
OPTICAL WIRELESS LAB

WASEDA UNIVERSITY

Slide 23 (bottom-left)

Next Generation FSO system

Transceiver design – optical antenna internal layout

beacon beam output windows
primary mirror
secondary mirror
FPM
collimation mirror
fiber connection port
QAPD/QPD

Courtesy of NiCT & Olympus

Design specifications

- Cassegrain configuration with 3 FFS mirrors (pri., sec. & coll.)
- Communication wavelength: 1550nm
- Beacon wavelength: 980nm
- Antenna aperture: 40mm
- Tracking mechanism:
  - Rough/coarse tracking: CCD camera
  - Fine tracking: QD
  - FPM for coupling the beam directly to the SMF

FFS: Free form surface
QD: Quad detector

23/69

Matsumoto
Mitsuji Laboratory
OPTICAL WIRELESS LAB

Slide 24 (bottom-right)

Next Generation FSO system cont.

Transceiver design – optical system components

Primary mirror
Collimation lense
BS1
BS2
to fiber port
QD
980 nm CCD
Secondary mirror
Top view

Primary mirror
Collimation lense
to fiber port
FPM
Secondary mirror
Side view

Optical system components showing optical path

- Design merits
  - Low coupling loss (less than 5 dB)
  - Achieve compact size and light weight (10 cm³, 1.6 kg)
  - 3 mirror design removes all major aberrations like spherical aberration, astigmatism, comma, chromatic aberration
- Demerits
  - Difficult to fabricate and alignment adjustment

24/69

Matsumoto
Mitsuji Laboratory
OPTICAL WIRELESS LAB

WASEDA UNIVERSITY

WASEDA UNIVERSITY

Next Generation FSO system cont.
Transceiver design – Fine tracking system

(b) Error signal generation for QPD/QAPD

On site measurement
Initial setting
(d) Typical intensity distribution at QPD/QAPD

X : azimuth
Y : elevation

Y-axis displacement
X-axis displacement
X-axis error signal
Y-axis error signal

Beacon light spot

Ethernet
Control/Monitor signal

APD Gain control
Sum
DSP
PC

QAPD
FPM
Error signal PID servo
Control/Monitor signal

(a) System block diagram

Beam (980nm)
Single mode fiber
HR coating (mirror)
AR coating (980nm)
Glass ferrule
Signal beam (1552nm)
AR coating (980/1552nm)
Focusing lens
(c) Fiber coupler

26/69

WASEDA UNIVERSITY

Next Generation FSO system cont.
Transceiver design – fluctuation suppression cont.

Fiber coupled signal power without FPM tracking

Received power (dBm)
Tracking OFF
Monitor output (V)
Time (msec)

Fiber coupled signal with FPM tracking

Received power (dBm)
Tracking ON
Monitor output (V)
Time (msec)

Tracking system reduces the pointing errors which suppresses the atmospheric induced scintillation effects.

28/69

WASEDA UNIVERSITY

Next Generation FSO system cont.
Transceiver design – Fine Pointing Mirror (FPM)

- Features
  - Non-resonant type two-axis Galvano mirror
  - Compact and integrated design with large tilt angle
  - Newly designed actuator enables high bandwidth and fine tilt control
  - Built-in angular sensor and actuator drivers

- Outline specifications
  - Tracking speed: 2 kHz
  - Mirror Φ : 2.3mm × 2.8mm
  - Tilt angle range: Az/El: ±6~8deg
  - Sensitivity of angular sensor : 140mV/deg
  - Tilt angle accuracy : < 0.02 m radian
  - Wave front accuracy of mirror : 0.04 λ rms

LD
Position sensitive detector
Following force spider
Two mini servos coil actuator
Collimating lens
Focusing lens
Mirror (Φ: 2.3mm × 2.8mm)

25/69

WASEDA UNIVERSITY

Next Generation FSO system cont.
Transceiver design – fluctuation suppression

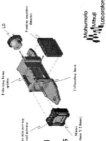

Pointing/tracking jitter suppression

Gain (dB)
Frequency (Hz)

More than 40dB suppression is achieved at the frequency less than 100 Hz.

Shock disturbance applied
10dB/div
1550 nm signal intensity
Tracking error in Az
Tracking error in El
74μrad/div

A shock disturbance is applied to the antenna module, but, the signal receive intensity remains unchanged.

27/69

WASEDA UNIVERSITY

Experimental field

Bldg. 14 Waseda University
Nishi Waseda Campus

1 km

Bldg. 55 Waseda University
Okubo Campus

Source: Google earth

Matsumoto
Matsuji
Laboratory
OPTICAL & WIRELESS LAB

Satellite view of the test area 29/69

WASEDA UNIVERSITY

WASEDA UNIVERSITY

Early experiment setup
Stratospheric platform experiment

Before

Just tracking

An airship behind a light antenna on ground.

Photograph of a prototype optical antenna for stratospheric platform experiment

Monitored image of the airship by CCD of ground station.

Matsumoto
Matsuji
Laboratory
OPTICAL & WIRELESS LAB

30/69

WASEDA UNIVERSITY

WASEDA UNIVERSITY

New experiment setup
Terrestrial communication experiment

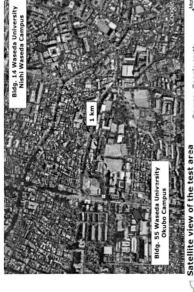

Weather monitor

Scintillation measurement antenna

Interface PC

Test antenna

CCD and FPM tracking control circuit

Beacon laser source

Devices setup

Test antenna

Matsumoto
Matsuji
Laboratory
OPTICAL & WIRELESS LAB

WASEDA UNIVERSITY

WASEDA UNIVERSITY

Experiment setup cont.

Okubo Campus

Bldg 14 rooftop

Scintillation measurement antenna

Test antenna

Nishi Waseda Campus

Bldg 55 rooftop

Lightwave clock/data receive & transmitter

Fiber amplifier

Weather data recording PC

Power meter

TV monitor

Remote adjustment & monitor PC

BERT

Scintillation data recording PC

Experimental hardware setup

Data collected include:
- Weather – temp., visibility, precipitation & fog
- BER
- Scintillation & optical attenuation
- Received optical power

Building rooftop setup

Matsumoto
Matsuji
Laboratory
OPTICAL & WIRELESS LAB

32/69

WASEDA UNIVERSITY

WASEDA UNIVERSITY

Experiment setup cont.

Measure and record scintillation and beam intensity variations caused by angle-of-arrival (AOA) fluctuations from atmospheric turbulences

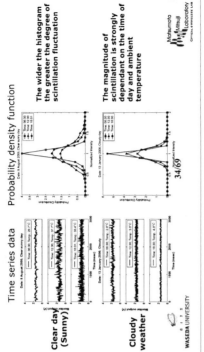

Bldg. 55
Waseda University
Okubo Campus

Scintillation attenuation measurement antenna

Test antenna

Actual rooftop installation

Attenuation & scintillation measurement antenna

Electrical signal

800nm — Photo detector → A/D converter → PC

1550nm — SMF — O/E receiver

Optical antenna under investigation (measurement of AOA fluctuations)

Measurement setup block diagram

33/69

WASEDA UNIVERSITY

Results: Scintillation effects

Comparison figures of intensity fluctuation as a result of scintillation

Time series data   Probability density function

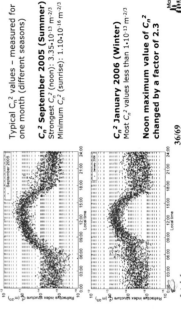

Clear day (Sunny)

Cloudy weather

The wider the histogram the greater the degree of scintillation fluctuation

The magnitude of scintillation is strongly dependant on the time of day and ambient temperature

34/69

WASEDA UNIVERSITY

Results: Scintillation effects cont.

Comparison of degree of scintillation for difference seasons

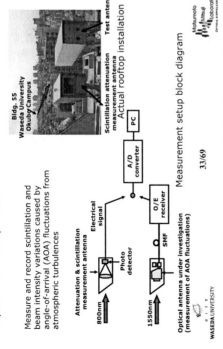

Hot season (Summer)

Cold season (Winter)

The degree of scintillation fluctuation varies with the time of day and ambient temperature and is more pronounced in the hot season as compared to the cold season.

Clear diurnal and seasonal variations are demonstrated.

35/69

WASEDA UNIVERSITY

Results: Atmospheric effects

Refractive-index structure constant parameter, C_n^2, is the most critical parameter along the propagation path in characterizing the effects of atmospheric turbulence

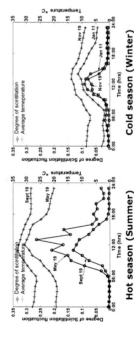

Typical C_n^2 values – measured for one month (different seasons)

C_n^2 September 2005 (Summer)
Strongest C_n^2 (noon): $3.35 \cdot 10^{-13}$ m$^{-2/3}$
Minimum C_n^2 (sunrise): $1.10 \cdot 10^{-16}$ m$^{-2/3}$

C_n^2 January 2006 (Winter)
Most C_n^2 values less than $1 \cdot 10^{-13}$ m$^{-2/3}$

Noon maximum value of C_n^2 changed by a factor of 2.3

36/69

WASEDA UNIVERSITY

Performance evaluation

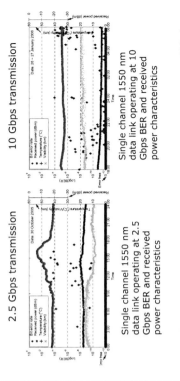

Experiments

- BER and received power characteristics
- WDM transmission characteristics
- Application data performance characteristics

Bldg. 55 Okubo Campus   Bldg. 14 Mishu Waseda Campus

Next Generation FSO communication system performance evaluation setup

Matsumoto Mitsuji Laboratory
OPTICAL & WIRELESS LAB

WASEDA UNIVERSITY

37/69

WASEDA UNIVERSITY

Results: BER and received power characteristics

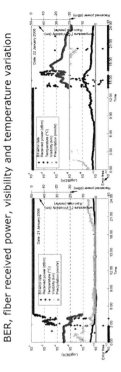

2.5 Gbps transmission   10 Gbps transmission

Single channel 1550 nm data link operating at 2.5 Gbps BER and received power characteristics

Single channel 1550 nm data link operating at 10 Gbps BER and received power characteristics

Matsumoto Mitsuji Laboratory
OPTICAL & WIRELESS LAB

WASEDA UNIVERSITY

38/69

WASEDA UNIVERSITY

Results: BER and received power characteristics cont.

BER, fiber received power, visibility and temperature variation

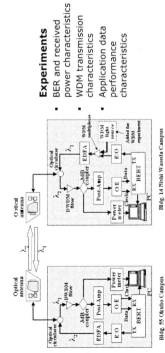

Single channel 1550 nm data link operating at 2.5 Gbps BER and received power characteristics under strong atmospheric turbulence

Single channel 1550 nm data link operating at 2.5 Gbps BER and received power characteristics under rain event

Matsumoto Mitsuji Laboratory
OPTICAL & WIRELESS LAB

WASEDA UNIVERSITY

39/69

WASEDA UNIVERSITY

Results: BER and received power characteristics cont.

BER, fiber received power, visibility and temperature variation

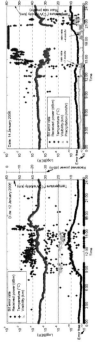

Single channel 1550 nm data link operating at 2.5 Gbps BER and received power characteristics under snow event

Matsumoto Mitsuji Laboratory
OPTICAL & WIRELESS LAB

WASEDA UNIVERSITY

40/69

WASEDA UNIVERSITY

Results: WDM transmission characteristics

Four channel 1550 nm data link operating at 2.5 Gbps WDM

WDM received signal spectrum

BER and received power characteristics

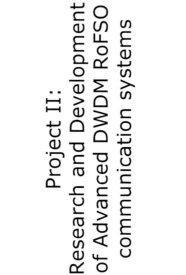

ITU Grid 100GHz spacing

1549.3 nm 1550.1 nm 1550.9 nm 1551.7 nm

- 2.5 Gbps X 4 channels with output power 100mW/wavelength
- Stable communication was achieved with no fluctuation or interference between wavelengths

42/69

WASEDA UNIVERSITY

Project II:
Research and Development
of Advanced DWDM RoFSO
communication systems

44/69

WASEDA UNIVERSITY

Results: Eye pattern characteristics

Effect of strong atmospheric turbulence

After transmission (worst case)

After transmission (typical case)

After transmission

Before transmission

2.5Gbps transmission

10 Gbps transmission

Most of the key wave shape parameters are within acceptable tolerance

41/69

WASEDA UNIVERSITY

Summary

- Successfully demonstrated an next generation FSO system capable of offering stable and reliable transmission at 10 Gbps **in the absence of severe weather conditions.**

- Measured, characterized and quantified the factors which influence the performance of the next generation FSO system in our deployment environment in particular atmospheric turbulence and weather related effects.

- The information is used in design optimization, evaluation, prediction and comparison of performance of next generation FSO system in operational environment.

43/69

WASEDA UNIVERSITY

Contents

- Introduction
  - Project details
- Overview of Advanced DWDM RoFSO
  - Research objectives
- Next Generation FSO system
  - Transceiver design and performance
- FSO system experiment performance results and analysis
- Summary

WASEDA UNIVERSITY

Introduction
Project details

- **Period:**
  - April 2006 ~ March 2009 (3 years)
- **Collaborating entities**
  - Waseda University
  - Osaka University
  - Sponsoring organization NiCT

WASEDA UNIVERSITY

Overview of DWDM RoFSO Link research

I. Development of an Advanced DWDM RoFSO System
- Transparent and broadband connection between free-space and optical fiber
- DWDM technologies for multiplexing of various wireless communications and broadcasting services

II. Development of Seamless Connecting Equipments between RoF, RoFSO and Wireless Systems
- Wireless service zone design
- Total link design through RoF, RoFSO, and Radio Links

III. Long-term Demonstrative Measurements
- Pragmatic examination of advanced ReFSO link system
- Investigation of scintillation influence on various types of wireless services transported using the RoFSO system.

WASEDA UNIVERSITY

Overview of DWDM RoFSO Link research

Key research objectives:

- Develop an advanced DWDM RoFSO communication system for transmitting heterogeneous wireless signal.
- Conduct system performance evaluation in actual deployment scenarios.
- Establish key issues which are significant in the design, evaluation, prediction and comparison of RoFSO system performance in operational environment.
- Conduct long term studies to characterize factors which influence performance of RoFSO system

WASEDA UNIVERSITY

New RoFSO system experiment setup

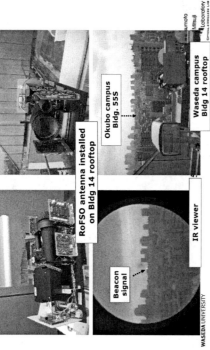

WASEDA UNIVERSITY

WASEDA UNIVERSITY

New RoFSO system experiment setup cont.

RoFSO antenna installed on Bldg 14 rooftop

Okubo campus Bldg. 55S

Waseda campus Bldg 14 rooftop

IR viewer

Beacon signal

WASEDA UNIVERSITY

WASEDA UNIVERSITY

New RoFSO system experiment setup

Main transmit and receive aperture

Si PIN QPD for coarse tracking using beacon signal

BS1

BS2

SMF

collimator

FPM (Fine Pointing Mirror)

LED

InGaAs PIN QPD for fine tracking

Beacon signal transmit aperture

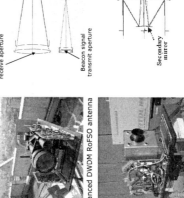

Post EDFA   Digital mobile radio transmitter tester (Anritsu MS8609A)

DWDM D-MUX

Optical source

Boost EDFA

Optical power meter (Agilent 8163A)

RoFSO antenna tracking adjustment and monitoring PC

Bit Error Rate Tester (Advantest D3371)

Atmospheric turbulence effects recording PC

Main transmit and receive aperture

Rough tracking beacon projection aperture

Bldg. 14 Nishi Waseda campus

Atmospheric effects measurement antenna

RF-FSO antenna

RofSO antenna

Weather measurement device

Bldg. 55S Okubo campus

WASEDA UNIVERSITY

WASEDA UNIVERSITY

Antenna comparisons: DWDM RoFSO and Next Generation FSO antenna

Main transmit and receive aperture

Si PIN QPD for coarse tracking using beacon signal

BS1

BS2

SMF

collimator

FPM (Fine Pointing Mirror)

LD

InGaAs PIN QPD for fine tracking

Beacon signal transmit aperture

Primary mirror

Collimation lense

BS1

980 nm CCD

BS2 to fiber port

QD

Secondary mirror

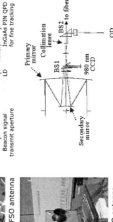

Advanced DWDM RoFSO antenna

Next Generation FSO antenna

WASEDA UNIVERSITY

WASEDA UNIVERSITY

New RoFSO system experiment

Characteristics of FSO antennas used in the experiment

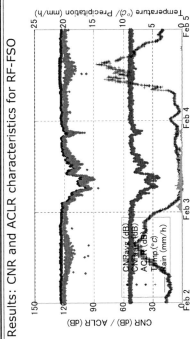

Parameter	Specification	
	RF-FSO	RoFSO
Operating wavelength	785 nm	1550 nm
Transmit power	14 mW (11.5 dBm)	30 mW (14.8 dBm)
Antenna aperture	100 mm	80 mm
Coupling loss	3 dB	5 dB
Beam divergence	± 0.5 mrad	± 47.3 μrad
Frequency range of operation	450 kHz – 420 MHz	~ 5 GHz
Fiber coupling technique	OE/EO conversion is necessary	Direct coupling using FPM
WDM	Not possible	Possible (20 dBm/wave)
Tracking method	Automatic	Automatic using QPD Rough: 850 nm Fine: 1550 nm

54/69

WASEDA UNIVERSITY

Matsumoto Mitsuji Laboratory · OPTICAL & WIRELESS LAB

WASEDA UNIVERSITY

New RoFSO system experiment setup diagram

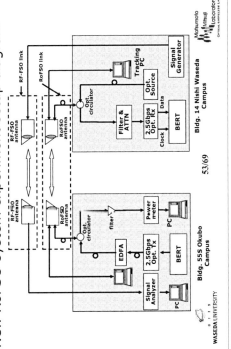

Bldg. 55S Okubo Campus

Bldg. 14 Nishi Waseda Campus

53/69

WASEDA UNIVERSITY

Matsumoto Mitsuji Laboratory · OPTICAL & WIRELESS LAB

WASEDA UNIVERSITY

Results: CNR and ACLR characteristics for RF-FSO

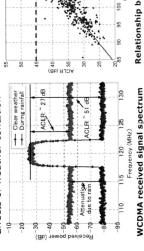

RF signal transmission characteristics measured using RF-FSO system

56/69

WASEDA UNIVERSITY

Matsumoto Mitsuji Laboratory · OPTICAL & WIRELESS LAB

WASEDA UNIVERSITY

Results: CNR and ACLR characteristics for RF-FSO cont.

Effects of weather condition

WCDMA received signal spectrum

Relationship between CNR and ACLR

WCDMA: Wideband Code Division Multiple Access
CNR: Carrier to Noise Ratio
ACLR: Adjacent Channel Leakage Ratio (a quality metric parameter for WCDMA signal transmission)

55/69

WASEDA UNIVERSITY

Matsumoto Mitsuji Laboratory · OPTICAL & WIRELESS LAB

WASEDA UNIVERSITY

Results: BER and received power characteristics
RoFSO system

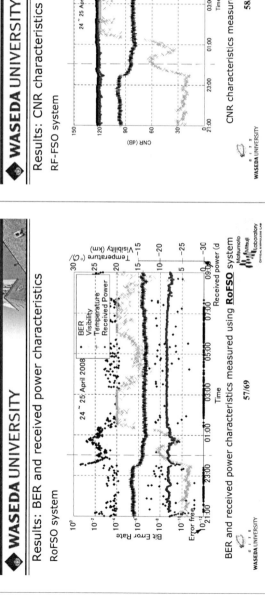

BER and received power characteristics measured using **RoFSO** system

57/69

WASEDA UNIVERSITY

WASEDA UNIVERSITY

Results: CNR characteristics
RF-FSO system

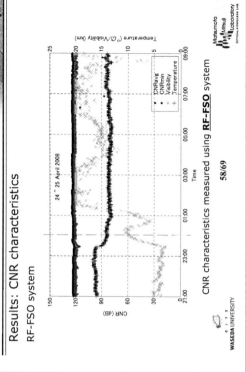

CNR characteristics measured using **RF-FSO** system

58/69

WASEDA UNIVERSITY

WASEDA UNIVERSITY

Results: ACLR and optical received power measurement
RoFSO system

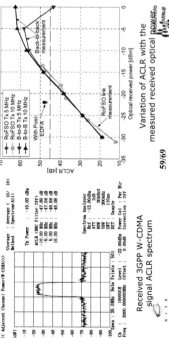

Received 3GPP W-CDMA
signal ACLR spectrum

Variation of ACLR with the
measured received optical power

59/69

WASEDA UNIVERSITY

WASEDA UNIVERSITY

Results: BER and optical received power measurement
RoFSO system

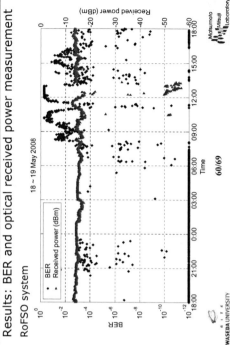

60/69

WASEDA UNIVERSITY

WASEDA UNIVERSITY

Summary

- Demonstrated initial results on performance of RoFSO system in actual deployment environment.
- Highlighted the implementation issues related to RF signal transmission over FSO channels.
- Further work on transmission of multiple RF signals using DWDM technology using the RoFSO system are ongoing

61/69

WASEDA UNIVERSITY

WASEDA UNIVERSITY

Introduction to GITS and MMLab research activities

62/69

WASEDA UNIVERSITY

WASEDA UNIVERSITY

Introduction to GITS and MMLab

Graduate School of Global Information and Telecommunications Studies (GITS)

Courses offered in GITS
- Computer systems and Network Engineering field
- Multimedia Science and Arts field
- Info-telecom Socio-Economics, Network Business and Policy field

GITS main campus Honjo Saitama

*GITS – Graduate School of Global Information and Telecommunication Studies

63/69

WASEDA UNIVERSITY

WASEDA UNIVERSITY

Introduction to GITS and MMLab cont.

Matsumoto Lab (MMLab)

Members
- Prof. Matsumoto
- Visiting professors
- Visiting researchers
- Doctor degree candidates
- Masters degree candidates
- Research students
- Exchange program students

MMLab locations
- Honjo Campus
- Waseda Campus Bldg. 29-7
- Okubo Campus Bldg. 55S

Prof. Mitsuji Matsumoto

http://evangelion.ml.giti.waseda.ac.jp/

64/69

WASEDA UNIVERSITY

WASEDA UNIVERSITY

Introduction to GITS and MMLab cont.

16.02.2007

08.02.2008

WASEDA UNIVERSITY

66/69

WASEDA UNIVERSITY

Introduction to GITS and MMLab cont.

Fellowships for visiting researchers and professors

- Ministry of Education, Culture, Sports, Science and Technology (Monbukagakusho)
- JSPS
  - http://www.jsps.go.jp/english/e-fellow/fellow.html
- NiCT
- Waseda Univ.

For further information please contact Prof. Mitsuji Matsumoto
Email: mmatsumoto@waseda.jp
or visit MMLab webpage at:
http://evangelion.ml.giti.waseda.ac.jp/

WASEDA UNIVERSITY

68/69

WASEDA UNIVERSITY

Introduction to GITS and MMLab cont.

MMLab main research topics

- Free-Space Optics (FSO) group
- Personal Area Networks (PAN) group
  - IrDA
  - Visible light communication
- Wireless Sensor Network (WSN) group
  - Location management using RFID
- e-Applications group
  - Network architecture
  - E-Learning
- Interdisciplinary research group

WASEDA UNIVERSITY

65/69

WASEDA UNIVERSITY

Introduction to GITS and MMLab cont.

Various countries with members in MMLab

WASEDA UNIVERSITY

67/69

WASEDA UNIVERSITY

Afternoon presentations

- Chen Yanru

  Title: P2P Solution of Large Data Video Processing System

- Christian Sousa

  Title: Fluent Interaction with Multi-Touch in Traditional GUI Environments

69/69

WASEDA UNIVERSITY

Thank you for your attention!

kazaura@aoni.waseda.jp